Mushrooms
and
Toadstools

a field guide

Geoffrey Kibby

Illustrated by
Sean Milne

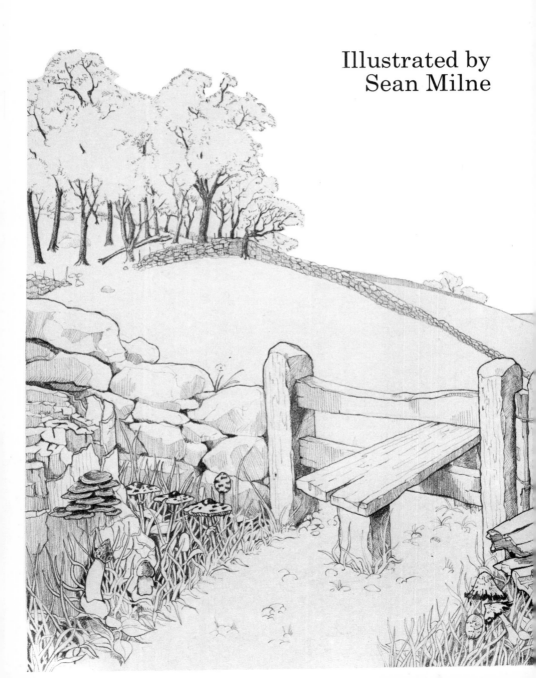

Mushrooms
and
Toadstools

a field guide

Oxford   New York   Toronto   Melbourne
OXFORD UNIVERSITY PRESS 1979

Editor: Bill MacKeith
Production: Alan Peebles
Editorial Director: Ben Lenthall
Design: Adrian Hodgkins

The publishers gratefully acknowledge
the assistance of Dr F. B. Hora

Oxford University Press
Walton Street, Oxford OX2 6DP
OXFORD LONDON GLASGOW MELBOURNE WELLINGTON
NAIROBI DAR ES SALAAM CAPE TOWN KUALA LUMPUR
SINGAPORE JAKARTA HONG KONG TOKYO DELHI BOMBAY
CALCUTTA MADRAS KARACHI

First published 1979

Planned and produced by Elsevier
Publishing Projects (UK) Ltd, Oxford
© 1979 Elsevier Publishing Projects SA, Lausanne

ISBN 0 19 217688 9
ISBN 0 19 286004 6 Pbk

Origination by Art Color Offset, Rome, Italy
Printed by Brepols, Turnhout, Belgium

PREFACE

There has been a great surge of interest in mushrooms and toadstools since the late 1960s, as more and more people have become aware of the range of colours and shapes, many of them striking, exhibited by this group of organisms. Unfortunately this greatly increased interest has not always been best served by the books which have attempted to satisfy it.

It is therefore a pleasure to introduce Sean Milne's accurately observed illustrations, which allow rapid identification in the field and are also beautiful paintings in their own right, and Geoffrey Kibby's text which is most useful and clearly written yet compact. Some of the species included have not previously been depicted in English language books, or indeed anywhere. The key to genera, with its margin illustrations and colour coordinates, is commendably easy for the amateur to use.

The publishers, author and artist have recognized the difficulties in treating the topic of larger fungi, especially as it is possible to give only a selection of the thousands of species known to occur in Europe, and I commend their selection and presentation. This field guide is a truly valuable addition to our mycological literature. It will be of use to a whole spectrum of natural historians and should become a standard reference work for the amateur collector.

Roy Watling
Senior Scientific Officer
Royal Botanic Garden
Edinburgh

CONTENTS

ERRATA

p. 77: the numerals 2 and 3 should be transposed

p. 129: the numerals should read 1 (top left),
2 (top right), 3 (bottom left), 4 (bottom right)

Fungi are a very distinct group of organisms. They are now generally considered as a separate kingdom of the living world, being as different from the other "plant" kingdoms – the bacteria, algae and green plants – as they are from the animal kingdom. Some 100,000 species of fungi have been described, while probably the same number await discovery. The vast majority of fungi are minute: many only visible to the naked eye as mats of cotton-like threads, such as the blue mould of *Penicillium* often found on stale bread. It is impossible to identify these fungi without the aid of a microscope. For most of us, however, the word "fungus" conjures up an image of the mushrooms and toadstools that are found growing in our woods and fields. It is these "large fungi" with which this book is concerned and which are illustrated in the plates.

Together with the bacteria, the fungi are the "decomposers" of our environment and they are just as important as the green plants which are the "producers". Fungi lack the green pigment, chlorophyll, found in the green plants and some algae. It is this pigment that is essential in converting the energy of sunlight directly into a form that is available for use by these higher plants. Because fungi do not have this pigment they must obtain the energy required for their life-giving processes indirectly from other sources. Fungi basically utilize the organic materials produced by other plants and animals in two ways. They may "feed" on the dead or decaying organisms or the products thereof, whether this be the dead trunk of a tree, the decomposed remains of a plant in the soil, a dead insect, or man-made products derived from plants and animals, such as foodstuffs and clothes. This form of obtaining nutrients is known as *saprophytism* and the fungi are hence *saprophytes*. The other main form of nutrition is by way of *parasitism*. Whereas saprophytism is a highly beneficial side to the life of fungi in that nutrients are cycled within the environment without harm to other life, parasitism represents the "dark" side of fungal life styles – at least from Man's point of view. These fungi live on living plants and animals, obtaining all the nutrients they require from the living host. In many cases the result will be the death of the host, but not normally until the fungus has completed its life cycle. Some fungi will in fact first feed parasitically on a host but, having killed it, will then feed saprophytically on the remains. It is with this side of fungal nutrition that Man most comes into contention. Fungi are the most serious pests of his crops and garden plants. It comes as no surprise that an attack of a microscopic fungus can kill a mammoth 100-year-old tree in a matter of a single year.

In recent years it has been found that many fungi in fact live in close harmony with the roots of living plants, without either partner suffering from the association and both normally benefiting. These symbiotic associations are known as *mycorrhizae* (fungus roots) and it is now known that most families of flowering plants form these associations with fungi. Without the fungus the vigour of the "host" plant suffers and certainly growth would be retarded; in some instances (including many orchids) the fungus association is essential if any growth is to occur at all. Most of these mycorrhizal fungi are mushrooms or toadstools and few people realize that without the fungus, which is only visible to the eye for a short time each year as a large fruit-body, many of our woods and forests would not exist in the form they do today.

The Structure of Larger Fungi

In the vast majority of fungi the basic unit of construction is the same: microscopic cotton-like threads called *hyphae*. The term *mycelium* is used to describe the complete system of hyphae that makes up each individual fungus. The mycelium forms the basic vegetative structure, each hypha penetrating and branching throughout the growing medium (e.g. humus, wood etc.), absorbing the required nutrients over its whole surface. Individual hyphae or parts of the mycelium become highly modified in the fruit-bodies, but even in the most complex "toadstool" the basic hyphal structure can still be recognized when viewed through a microscope. Many people believe that when they observe a mass of toadstools in autumn these are the only parts of the fungus that ever form. In fact all year round masses of hyphae are to be found growing in the surface layers of the soil and only at the time of reproduction do the fruit-bodies we all recognize develop and become visible.

The function of the fruit-bodies is to produce spores, which are dispersed and will germinate to produce new mycelial colonies. The form of the fruit-body is the basis for classifying fungi, and in the larger fungi described in this book the fruit-body characters are used to identify fungi right down to the species level.

Of the more than 10,000 fungus species in Britain alone some 2,500–3,000 may be described as "larger fungi" – those which are easily visible or can be handled. Many of these have very distinctive shapes and have over the years acquired such common names as puffballs, fairy clubs and earthstars (see the Index); many of these names are to be found in almost worldwide usage, with but little variation.

Some confusion arises because there is no clear difference between "mushrooms" and "toadstools" – both are words of convenience meaning different things to different people. In Britain "mushroom" may refer popularly only to the single cultivated species of the genus *Agaricus* sold in shops, or it may, as in this book, extend to representatives (not all edible) of half a dozen genera, including *all* species of *Agaricus* ("true" mushrooms). In other parts of Europe one term (e.g. *"champignon"*) often covers *all* fungi, both microscopic and the large numbers of fungi that are eaten and enjoyed, and in North America most edible fungi are referred to, by mycologists at least, as mushrooms. "Toadstool" generally implies inedibility or the fear of *poisonous* qualities; in this sense it is sometimes used to describe any larger fungus with an umbrella-shaped cap on a central stem, other than *Agaricus brunnescens* (=*A. bisporus*), the Cultivated Mushroom.

The whole fungal kingdom can be divided into five or more subdivisions, but only those belonging to the subdivisions Ascomycotina and Basidiomycotina grow to sufficient size to be recognized as larger fungi. However it must be noted that not all ascomycetes and basidiomycetes are larger fungi; many are microscopic and are hence not included here. Good examples of microscopic basidiomycete fungi are the rusts and smuts, serious pests of our crops. But, whatever the size, the reproductive structures for all members of one of these subdivisions is basically the same.

BASIDIOMYCOTINA

The majority (up to plate 93) of species described in this book are basidiomycetes (subdivision Basidiomycotina). This group includes the mushrooms, toadstools, puffballs, stinkhorns, and earthstars. The unifying feature of the members of this group is the nature of the spore-producing cells. (It must be noted that although the fungi dealt with here are "larger" fungi, the spore-producing apparatus is microscopic and cannot be seen with the naked eye.) The spores are produced by a club-shaped cell called the *basidium* and hence the spores are called basidiospores.

Within the basidiomycetes the position of these spore-producing cells varies and the further division of these fungi into smaller groups is based on this variation: the Hymenomycetes and Gasteromycetes. In the Hymenomycetes the fertile layer or *hymenium*, which contains the basidia, lines the surface of, for example, the gills of a gill fungus and the pores of a bracket fungus. At maturity of the fruit-body the hymenium is in contact with the outside air and mature spores are dispersed from the hymenium directly into the atmosphere. The spores are ejected violently into the space between the gills or within the pores. In the Gasteromycetes (for example the puffballs and earthstars) the basidia are formed in the internal tissues of the fruit-body and when mature are released within the "skin" of that body. Before the spores can be released to the atmosphere, and become dispersed, the outer wall must be broken, for example simply by disintegration or by the force of raindrops falling on the outer surface, which projects the spores into the atmosphere.

One group of Gasteromycetes, the stinkhorns (order Phallales), relies upon insects to disperse its spores: the foul-smelling spore-mass is quickly eaten by flies, which apparently pass the spores unharmed through their digestive tracts to later germinate. Another group of Gasteromycetes has become subterranean. These are the false truffles (order Hymeno-

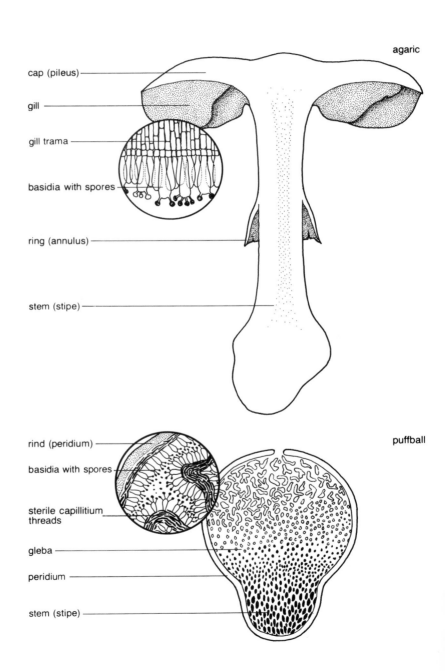

agaric

cap (pileus)

gill

gill trama

basidia with spores

ring (annulus)

stem (stipe)

puffball

rind (peridium)

basidia with spores

sterile capillitium threads

gleba

peridium

stem (stipe)

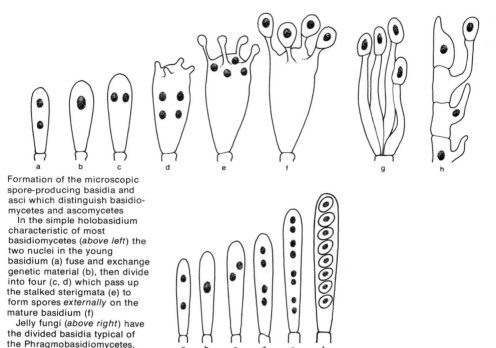

Formation of the microscopic spore-producing basidia and asci which distinguish basidiomycetes and ascomycetes

In the simple holobasidium characteristic of most basidiomycetes (*above left*) the two nuclei in the young basidium (a) fuse and exchange genetic material (b), then divide into four (c, d) which pass up the stalked sterigmata (e) to form spores *externally* on the mature basidium (f)

Jelly fungi (*above right*) have the divided basidia typical of the Phragmobasidiomycetes. The two shown are (g) *Tremella* and (h) *Auricularia*

In ascomycetes (*below*) the spores develop entirely *within* the ascus. The two nuclei within the young ascus fuse (a, b) and split into usually eight (c, d, e) before maturing

Left: structure in the larger basidiomycetes (Basidiomycotina)
The agaric has external basidia lining the gills. In other basidiomycetes the basidia may line tubes (brackets and boletes), spines (spine, tooth or hedgehog fungi), wrinkles, "veins" or flat surfaces.

Puffballs have their spore-carrying basidia *inside* the fruit-body, as do most other Gasteromycetes; the stinkhorns and bird's nest fungi are notable exceptions

gastrales), which rely on small mammals to dig them up and so disperse the spores.

The form of each basidial cell is also used as a means of splitting the Hymenomycetes into two groups. In one form each basidial cell bears four (sometimes two) stalks called sterigmata, from each of which a single spore is formed. This type of basidium is known as a holobasidium and hence the group is called the Holobasidiomycetes, which includes the gill fungi and bracket fungi. In the other group, each basidial cell is divided into four compartments by cross walls (septa) from each of which a spore-producing sterigma forms. This type of basidium is known as a phragmobasidium; the group Phragmobasidiomycetes includes the jelly fungi.

ASCOMYCOTINA

The other main group of fungi that concerns us here is the Ascomycotina. The vast majority of these are small and are not described in this book. But some are quite large and easily identified (see pp. 238–49). They include the cup fungi, morels, earthtongues and truffles (all disc fungi or Discomycetes) and the flask fungi (Pyrenomycetes). In the asco-

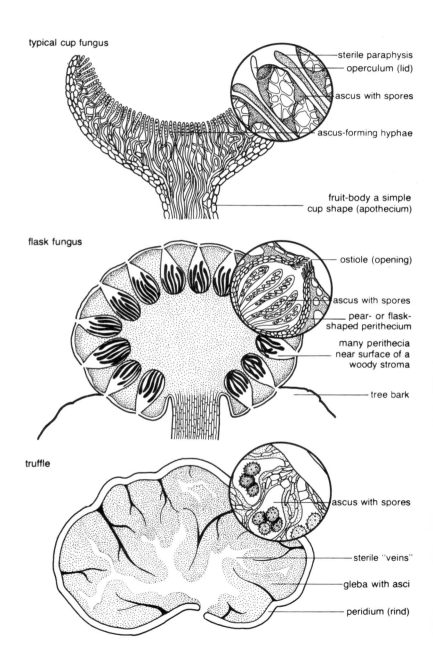

typical cup fungus

sterile paraphysis
operculum (lid)
ascus with spores
ascus-forming hyphae

fruit-body a simple
cup shape (apothecium)

flask fungus

ostiole (opening)

ascus with spores
pear- or flask-
shaped perithecium

many perithecia
near surface of a
woody stroma

tree bark

truffle

ascus with spores

sterile "veins"

gleba with asci

peridium (rind)

Those larger fungi in which the spores develop in asci (members of the subdivision Ascomycotina, or ascomycetes) fall into three main groups, illustrated here

In a typical cup fungus (order Pezizales) the fertile layer is exposed; not all cup fungi form simple cups – the morels for instance form complex sponge-like caps on stalks

In the flask fungi (Pyrenomycetes) the asci are grouped into "flasks" (perithecia); in larger species like the one illustrated many perithecia together form a stroma, which is usually hard and woody and often black

In the truffles (Tuberales) the asci are completely enclosed in the underground fruit-bodies. A third group within the same class (Discomycetes) is the earthtongues and their allies (order Helotiales), with minute cup- or disc-shaped apothecia and asci without a lid (operculum)

mycetes the specialized spore-producing cell is called an ascus and is characteristically sausage-shaped. Each ascus produces within it (note that spores are not produced *outside* the cell on stalks as in the basidiomycetes) eight spores, which are then dispersed through the tip of the ascus. The overall form of the fruit-body that produces the asci varies and is characteristic of the different groups. In the Discomycetes the asci are produced in a hymenium that lines the inner surface of a simple flattened or cup-shaped disc (apothecium). In the morels the cup is folded and convoluted like a sponge, with the hymenium lining the pockets, while in the truffles the cup is so convoluted and infolded that the hymenium has become totally enclosed and is subterranean. In the Pyrenomycetes the asci are produced in minute pear-shaped structures (perithecia) which open by a small pore (ostiole) to the outside. Many perithecia may be gathered together to form a stroma, usually a hard woody mass with the pores of the perithecia opening at the surface.

Spore dispersal in the Ascomycotina takes place when the spores have reached maximum size, and fluid pressure within the ascus is high. Suddenly the tip of the ascus will either just split (inoperculate) or a special lid-like area (operculum) will flip up and the spores shoot out like bullets from a gun, travelling several centimetres. This can be seen with the naked eye and also heard! If a mature fruit-body of the larger cup fungi is disturbed all the asci which are at maximum tension will fire off together and release a cloud of thousands of spores which will be seen as a puff of "smoke" and heard as very quiet rustling. However, in the truffles (order Tuberales), as in the false truffles of the Basidiomycotina, dispersal is usually by animals, although some species still retain the forcible dispersal mechanism.

Both basidia and asci may be interspersed with other, sterile cells. These are generally rather long, narrow, sometimes hooked cells acting probably as "packing" to support and separate the spore-producing cells. In the Basidiomycotina some very specialized and often very large cells (cystidia) can be found scattered over the hymenium and other parts of the fruit-body. Often many times larger than the basidia and of different shapes, cystidia can be of great importance in identification of difficult species; their function is not certain but they are often rich in oily substances.

The Typical Agaric

Most of the fungi illustrated in this book are members of the Agaricales (mushrooms and toadstools). A closer look at the life and structure of a typical agaric will help to reveal those characters used in identification, how and in what form they arise in the fruit-body. An example of a highly evolved and complex toadstool is the Fly Agaric, *Amanita muscaria*, and this will serve as a model. All fruit-bodies begin as a tiny knob of tissue arising from the underground mycelium; at first little structure can be distinguished but soon the tissues begin to differentiate and form the separate structures of the *cap* (technically called the pileus) and the *stem* (stipe). If you cut a very small button mushroom in half you can see the features appearing.

In the Fly Agaric at the young stage there is a surrounding "veil" of protective tissue which remains intact until the expansion of the fruit-body. This is usually referred to as the *universal veil*. When the fungus has expanded, the veil, forced to tear under the strain, is usually found clinging to the cap as white, woolly fragments or warts and at the base of the stem as rough gutter-like bands. Before expansion takes place however the gills of the toadstool will have developed. These are the radiating, very thin, plate-like structures found below the cap of most agarics and easily seen on the cultivated mushroom of the shops (*Agaricus brunnescens* = *A. bisporus*). These gills grow downwards from the undersurface of the young cap. In the Fly Agaric they are protected by a second veil which stretches from the edge of the cap inwards to the stem and is called the *partial veil*. When the cap expands this will be forced to tear away from the cap margin and is left hanging from the stem as a *ring* (or annulus). Not all toadstools have these veils of tissue. In some only the partial veil is present and instead of being a sheet of thick tissue as in the Fly Agaric it is merely a fine, cobweb-like film called the *cortina*. The nature of the different

In the more highly specialized and evolved agarics various protective tissues cover the fruit-body and its constituent parts during its development. In the toadstool shown, an *Amanita* species, two veils of tissue are involved, one an outer enclosing bag, the universal veil, which ruptures as the fruit-body expands to leave a volva at the base and fragments on the cap, the other an inner partial veil covering the developing gills, which is pulled away as the cap opens to leave a ring on the stem

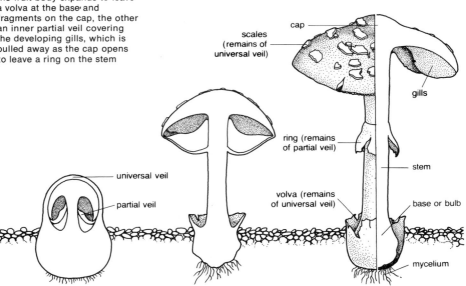

scales (remains of universal veil)

cap

gills

ring (remains of partial veil)

stem

universal veil

partial veil

volva (remains of universal veil)

base or bulb

mycelium

Above: the structure of the partial veil or ring usually falls into one of these categories: a) pendent; b) with a cortina or cobweb-like, often leaving a faint ring or ring zone; c) sheathing or stocking-like; d) thick, fleshy, turned-back

Below: in the genus *Amanita*, the volva is of paramount importance for correct identification: a) 2–4 hoop-like ridges (*A. pantherina*); b) irregular bands or ridges (*A. muscaria*); c) gutter-like or marginate bulb (*A. citrina; A. porphyria* with bag-like margin); d) bag-like with thin, irregularly torn margin (*A. phalloides, A. caesarea, A. vaginata* and allied species)

veils and the way they tear and are left attached to the fungus are often of great importance in the identification of species. This is best displayed in the members of the genus *Amanita*. In the Fly Agaric the universal veil is left as simple bands around the stem-base but if we look at *Amanita phalloides*, the Death Cap, it is seen to form a large, thick bag (volva) enclosing the stem. Other species of *Amanita* have slightly different volvas and ridges at the stem-base (see pp. 52–9) which are constant for each species and therefore help identification.

The gills of toadstools can vary enormously from genus to genus and between species in the same genus. Again these differences are of value in identification. The way the gills approach the stem and join on to it (the gill attachment) can be broadly classified as follows:

Free to *remote:* not connected to the stem, sometimes separated by a distinct gap or "collar".

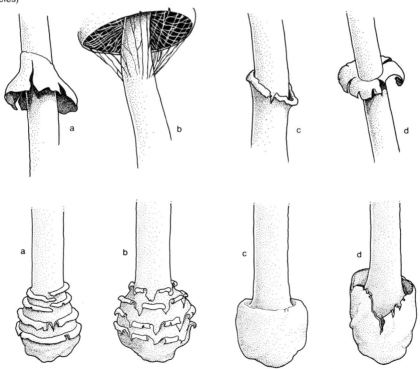

Adnexed to *adnate:* joining the stem for part (adnexed) or all (adnate) the depth of the gill.

Decurrent: joining the stem and "running down" it to a greater or lesser extent.

Sinuate or *emarginate:* the gill is notched just before joining the stem.

Combinations of these terms are often used to describe intermediate forms (e.g. adnate-decurrent, meaning the gill is broadly attached and tends to become slightly decurrent).

Not all larger fungi have gills; in the boletes their place is taken by closely packed tubes, but the description of pore-attachment to the stem remains the same.

The rapid expansion of the fruit-body (often a matter of a few hours) is exactly that, expansion as opposed to growth; most of the growth has already taken place in the previous days under-

Gill attachment
The way the gill approaches and joins onto the stem is often specific for certain genera and most species. Any gills will be of one of the following types (or a combination of two types, such as sinuate-decurrent); a) free; b) adnexed (just reaching stem); c) sinuate (with sudden notch or upward curve by stem = emarginate); d) decurrent (running down the stem to a greater or lesser extent); e) adnate (joined to stem by full depth of gill, but not running down the stem); f) sinuate, with decurrent "tooth" running down stem

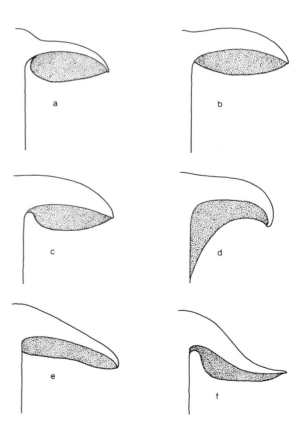

ground in the unexpanded "button". When conditions are right the button will start to absorb moisture, cells will expand, and under pressure the tissues will force upwards and outwards, tearing any veils that are present. During the expansion and often brief life of a fruit-body various mechanisms are employed to ensure the successful use of the gills and subsequent spore discharge. If the gills or tubes are not aligned exactly vertical in relation to gravity, then the spores, which are formed on the gills or tubes, will not be released into the air correctly and will hit and stick to adjoining gills or tubes instead of having a free, unhindered fall between them. If therefore the cap is for any reason misaligned the gills can correct this fault to some extent by repositioning, tilting to the vertical. The whole cap can also be realigned by the bending of the stem. This is easily demonstrated by placing an agaric on its side, preferably a young specimen just expanding: within a matter of an hour or two the stem will bend near the top at the apex so as to bring the cap horizontal; this, combined with minute gill movements, ensures successful spore discharge. The importance of this is obvious: stemmed fruit bodies growing out from a vertical bank (a common occurrence) can still disperse their spores successfully.

Gill colour varies considerably and is further altered by the colour of the spores when mature (the gill and spore colour are not necessarily the same). The main colour groups of spores are white to cream; shades of ochre to medium rust or deep brown; pink; cocoa or purple-brown to black. Other more unusual colours also occur rarely, such as red or blue-green, but they are exceptional. These colours form another useful adjunct to identification (see the Key), although it must be stressed it is an *artificial* one, as fungi with different spore colours are often closely related, and vice versa.

The textures of the fruit-body differ greatly so that there can be smooth, rough, velvety, hairy or sticky-capped (viscid) species and a great many other less easily defined textures. These are all of importance in identification and are frequently mentioned in the species descriptions and key later in the book. Consistency of the flesh may also help in identification, as for example with the russules and milk-caps (see pp. 168–83).

Although most agarics follow the basic shape of a cap surmounting a stem, these two features naturally vary in size, thickness, shape, etc. and several well-defined types are referred to. The cap can therefore be *subglobose*, expanding and then flattening; acutely pointed, which is usually referred to as *conical*; it can be *umbonate*, which means with a distinct bump or hump at the centre; dish-shaped caps are described as *depressed*, while if they are deeply funnel-shaped they are *infundibuliform*; the edge of the cap may be thinly grooved like the milled edge of a coin (*striate*) or more deeply furrowed (*sulcate*). The stem may be *equal* (i.e. parallel-sided); *clavate*, which means club-shaped, broadening toward the base; *bulbous* means with an abruptly swollen base; a *marginate* bulb refers to an abruptly defined bulb with a flattened upper edge. Other terms are explained in detail in the Glossary.

Useful and Harmful Fungi

All organisms, whether plant, animal or fungi, bring about changes in their environment simply by the normal processes of life, feeding, for example, excretion and respiration, and the results as far as man is concerned can often be highly beneficial or disastrous. Fungi are certainly one of the largest causes of trouble to man the world over. They attack his food crops, timbers, stored goods, clothing, buildings, his animals and even his own body. Even such apparently "non-edible" products as optical glass, paint, kerosene and plastics can be affected by fungi.

Perhaps the most readily observed fungal attacks are the moulds and mildews. The thick blackish-speckled mass on damp bread is a familiar sight; the fungus involved is *Rhizopus stolonifer*, the common bread mould, a member of a large group of fungi called the *Mucorales* or pin-moulds because of the pin-like heads which contain the spores. This same fungus can also be a serious cause of rotting in stored fruit, particularly apples,

Cap shapes
Cap shape is often important in identification, as it is usually constant within a species. The commonest shapes are shown here: a) conical; b) convex to convex-expanded; c) umbonate (with "humped" centre); d) campanulate (bell-like); e) expanded and flattened; f) funnel-like, infundibuliform; g) depressed at centre, umbilicate

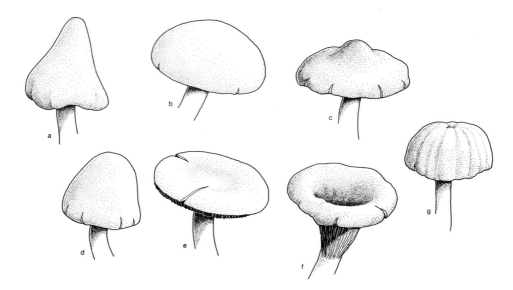

causing a soft brown circular patch which soon spreads until the whole apple is a soft, rotten mass. But the same species is used to man's advantage in commerce in the production of various chemicals such as cortisone. Many related species are also used to produce such important products as citric and oxalic acids and alcohol. This double character of being both a pest and a useful agent is very common among fungi. The familiar green and blue circular moulds caused by *Penicillium* species and their allies, although pests of stored foodstuffs such as citrus fruits, jams and jellies, apples, leather and fabrics, are also of course famous for their part in the historic development of antibiotics, including penicillin and its many derivatives. Other species of the genus produce the distinctive tastes in many of our foods: *P. roqueforti* flavours Roquefort cheese and *P. camemberti* does the same for Camembert. Penicillium moulds are also used in the production of Danish blue and Gorgonzola cheeses. Like the pin moulds they are also effective producers of various organic acids in commerce. The related genus *Aspergillus*, which includes some of the common black moulds, contains some of the most successful of all fungi: aspergilli can attack almost any damp substance, while some are serious causes of disease in livestock and occasionally in man, usually affecting the respiratory organs to cause aspergillosis. As with *Penicillium* they are also used in industry as fermenting agents and to produce acids.

The yeasts, simple, mainly unicellular fungi are a part of the subdivision Ascomycotina. They are distant relatives of the cup fungi illustrated in this book which produce spores in cells called asci. However in the yeasts this method of reproduction appears to be less frequent than asexual reproduction by budding or transverse division. It would be fair to say that the yeasts have had the greatest impact of any fungus on man and that life would be truly different without them; the key to their usefulness lies in their ability to ferment sugars. The alcohol industry, bakeries, and also some vitamin production depends on the action of yeasts in releasing either carbon dioxide or alcohol as they break down carbohydrates. When oxygen is present carbon dioxide is the waste product while in oxygen-deficient conditions alcohol is produced. Yeasts also impart special flavours to their byproducts and often contain high quantities of various complex substances such as the vitamin B-complex. Other yeasts are cultivated for direct consumption as food, being a rich source of protein and amino acids.

However as with other fungi, yeasts can also be pests, some causing troubles in the brewing industry and others in some foods by producing unpleasant flavours. Man can also suffer from the serious disease cryptococcosis caused by a yeast which affects either the respiratory system or more usually the nervous system when it is known as *Cryptococcus meningitis*.

Turning to our crops in the field, such as wheat, barley and other grasses, together with all the many other vegetables and

Stem types
Stem shape is often very distinctive in agarics and various terms describe the commoner types a) simple, cylindrical, equal (parallel-sided); b) bulbous (rather abruptly so in the example shown); c) marginately bulbous (with distinct ridge to upper edge of bulb); d) rooting, radicant; e) clavate (evenly swelling, club-shaped); f) lateral (from one edge of cap); g) eccentric (below cap but set to one side)

fruits grown by man, are all prey to a host of plant pathogens of fungal origin. The rusts and smuts are just such pathogens and both are placed by many in the class Basidiomycotina. The rusts (order Uredinales) comprise parasitic fungi which cause enormous losses of crops and which often have very complex life-histories involving different host plants, and various resting or overwintering states. Typical of the rusts is *Puccinia graminis* the Black or Stem Rust of wheat. This overwinters on wheat stubble and straw in the form of special spores called teleutospores, which germinate in spring. The resulting mycelium gives rise to normal basidiospores which when dispersed infect a secondary host, the barberry. A further type of spore, the aecidiospore, is then produced which is violently dispersed in great numbers, whereupon it infects fresh wheat plants which then produce the familiar speckled or "rust" symptom.

At this stage yet another type of spore, the uredospore, is produced which is not violently discharged but is easily dispersed by wind to infect other wheat plants, but not the barberry host. Uredospores and the resultant infections occur throughout the summer months, but as autumn approaches teleutospores are again produced, and so the cycle starts again. This cycle of *Puccinia graminis* is typical of many other rust species which also have two hosts. Some species, however, have only one host, such as the blackberry rust, *Phragmidium violaceum*.

Rusts cause serious damage in such various crops as coffee, conifers, asparagus, beans and carnations. All cost us money and time and a continual battle is being waged between man and fungus, with the production of new fungicides and the breeding of disease-resistant strains of food-plants.

The smuts (order Ustilaginales) are also important pathogens of plants but do not have such complex life-cycles as the rusts and never alternate from one plant to another. Again various cereals and grasses are affected as well as many other flowering plants and the infection is typically a dry, dusty, "smutty" coating. The spores are often produced in the reproductive organs of the host plant so that, for example, in the smut of false oats caused by *Ustilago avenae* the grain is replaced by smut spores.

The well-known ergot, *Claviceps purpurea*, which infects various cereal crops, is an ascomycete that has produced violent and extensive outbreaks of disease since historic times. It produces a black, pointed body that protrudes from the seed heads of grasses and cereals and is formed of matted hyphae bound together to form a mass called a sclerotium. This "ergot" contains several very *dangerous* and potent alkaloids which severely affect both animals and man, with varied symptoms which lead in many cases to death. Bread made from infected grain has in past ages wiped out very large numbers of people; thankfully, with modern harvesting and cleaning techniques as well as the use of fungicides, ergotism as it is called is now a rare occurrence.

Many other fungi cause diseases in man such as athlete's foot and ringworm, both of which are common and troublesome the world over, while others are much rarer and more restricted, although frequently very serious. These include blastomycosis and related forms which cause deeper, often chronic, infections of body tissue. Even toadstools are not above infecting man – *Schizophyllum commune*, a worldwide agaric species found on timber has, albeit very rarely, been found causing fungal infections in human beings.

The larger fungi are however much more familiarly known as destroyers of human property, especially in the notorious Dry Rot, *Serpula lacrymans*. This basidiomycete is a relative of the familiar bracket fungi of our woodlands, which is almost exclusively found in houses, forming spreading, highly invasive sheets of tissue which soon decay and soften wood, plaster etc. until the whole structure collapses, an all too common and expensive occurrence. Other toadstools are, of course, of use to man as food and conversely some poison him when mistaken for edible species, but these are discussed in detail in the following pages.

Fungi as Food

If you are a collector of fungi, sooner or later you may wish to eat some. Then the question of fungi as food will arise – just how safe are they, are they good to eat and do they contain any useful vitamins or protein? These are some of the questions being asked more than ever today when collecting wild foods is all the trend and so many books are encouraging you to try new and exotic foods. Firstly it must be made quite plain that there are poisonous fungi and that they can be, and sadly often are, fatal particularly in Europe and occasionally in America, although very rarely in Britain. Having said this it is only fair to state that by far the majority of the larger fungi are not poisonous and many are both edible and delicious. This being so one must find some guidelines for distinguishing the two and it is here that so many people go astray. *None* of the old folk traditions regarding wild fungi and how to "test" them for safety hold true – repeat NONE! Cooking fungi with a silver sixpence or spoon, to see if it turns black, peeling the cap, boiling in salt water, only using fungi in fields, only using white toadstools – are a few of the traditional (and very much alive even today) tests for safety. These tests simply will not work against all poisonous fungi. The Death Cap, *Amanita phalloides*, WILL PASS all these tests (even the last two on occasions) and still possibly kill you. The only safe and sure method is to be able to recognize the edible and poisonous species concerned by sight and never to eat anything which is in the least doubtful. Never take notice of helpful and well-meaning friends with a smattering of country lore, try instead to obtain identification from an expert until you are proficient yourself. Attend organized fungus forays in your area; most natural history societies run them or know of a local expert who can help.

The recognition of edible species is by no means a difficult task if certain very well known and highly valued species are chosen. There are half a dozen or more delicious species easily identified even by complete beginners and which are entirely safe and eaten the world over. None of these is a true mushroom of the genus *Agaricus* to which the shop mushroom belongs – the wild mushrooms can be surprisingly difficult for a beginner to identify and this has led in the past to accidents. The following species are all quite distinct from the ordinary mushroom and the *dangerous* toadstools in both shape and colour. Firstly and the undoubted king in the fungus world is *Boletus edulis* (see pp. 186–7) known as the Cep in France, Steinpilz in Germany, Penny Bun in some parts of England plus other names. This is widely considered the most delicious and useful of the edible fungi, an opinion with which the author would agree! It is a bolete, which means it has tubes underneath the cap instead of gills. This immediately rules out nearly all the dangerous toadstools which are almost entirely gilled species. Secondly come the Morels – the various species of *Morchella* (see pp. 238–9). These occur in the spring and look like a sponge on a stalk. Again very easy to recognize, they are considered by many, especially the Americans, as equal and possibly superior to the previous species. Next is the Chanterelle, *Cantherellus cibarius*, a beautiful egg-yellow to apricot toadstool with a top-shaped cap and blunt, ridgelike "gills", again a widely valued species sold all over Europe in the markets (see pp. 206–7). The Shaggy Ink-cap (pp. 76–7) is another unmistakable species with a very tall, cylindrical, shaggy-white cap soon turning to black inky liquid; picked when still young and solid it makes a delicately flavoured and tender delicacy stewed or baked. The Giant Puffball is exactly what its name suggests – a huge white ball-like fungus which when mature produces a powdery white spore-mass but when still young, can be cut into steaks and fried with very good results; it reaches 10–40 cm across or more (see pp. 232–3). The Blewits (pp. 128–9), *Lepista nuda* and *L. saeva* are a personal favourite and well liked by a great many mycophagists (as fungus eaters are known). (For a first tasting of Blewits a small quantity is recommended: a few people have an allergic reaction to these species.) Their lilac-violet colouring and very late (autumn to early winter) appearance make them again easy to recognize.

These six species would make a good selection to start with and are impossible to confuse with such things as the Death Cap. Many other edible fungi can also be recommended among which the best are the Parasols, *Macrolepiota procera* and *M. rhacodes*; the Black Trumpets, or Horn of Plenty, *Craterellus cornucopioides*, Oyster mushroom, *Pleurotus ostreatus*, St. George's Mushroom, *Calocybe gambosum*, many of the *Boletus* species, and finally the Beafsteak or Ox-tongue fungus, *Fistulina hepatica*. All these species are illustrated in this book and are worth trying if the opportunity arises.

Always eat a very small quantity of a new fungus at first – some people are allergic to many foods including fungi, others just find them indigestible. Do not leave picked fungi for long before preparing them for food or for preserves or placing in the freezer. Fungi inadvisably placed in a polythene bag will decompose fast. Only use fresh, young material for cooking although not so young as to be unrecognizable. To prepare fungi for cooking is very easy. Simply wipe the cap with a damp cloth; avoid soaking or too much washing – this makes them soggy. Do not bother to peel the caps, as this is entirely unnecessary and a waste of effort. Gills and pores may also be left on unless they are very soft or are damaged. The stem is usually removed if tough; otherwise chop or slice it along with the cap. Fungi may be oven-dried for storage and later use or even pickled as ketchup.

Any method of cooking used for green vegetables may be used for fungi – stewing, steaming, baking etc.– and fungi may also be treated as a meat dish by frying, grilling etc. Many species are delicious when stuffed with a spicy filling or used in omelettes. Use your imagination, and fungi can become a valuable addition to any meal. In addition, many cookery books give recipes for fungi and some volumes concentrate exclusively on such cooking.

Poisonous Fungi

Hopefully you will never experience the effects of eating a poisonous fungus, but it is still important (and interesting) to know what they are and how the poisons work. The number of really dangerous species of fungi is relatively small and restricted on the whole to one or two particular genera. However there are a number of other poisonous species which although perhaps not deadly differ widely in their effects on the human body and in the chemicals involved in the poisoning. Because of these differences it is possible broadly to classify those fungi by their poisons into a few basic groups.

Protoplasmic toxins – the deadly amanitas

Foremost amongst poisonous fungi must come the various species of the genus *Amanita*, in particular the infamous Death Cap, *Amanita phalloides* (pp. 52–3) and the less well known but equally deadly Destroying Angel, *A. virosa* (pp. 56–7). No other fungi have such complex or numerous toxins and this fact has made treatment in the past extremely difficult and uncertain. The poisons in *Amanita* fall into two main groups, the phallotoxins and the amatoxins, the two groups differing in their effects on the body and the time involved in their actions. In the Death Cap there are believed to be six related phallotoxins and five or more amatoxins. The symptoms produced by these toxins are many and variable. After eating a meal of Death Cap (which incidentally tastes quite good) the first effects are often not felt for some 5–30 hours, thus making stomach-pumps of little use, as the toxins have been well distributed by the blood-stream. Firstly the victim will suffer from abdominal pains, diarrhoea and vomiting. This will increase until the victim is in a state of shock from fluid losses, exhaustion, etc., and if treatment is not carried out death soon ensues. If treatment is available these symptoms are usually soon combated and the patient may seem to revive for a day or two, but then quite suddenly will suffer a relapse where the pulse is weak, blood pressure drops drastically and there may be hallucinations. Finally death occurs and an autopsy shows major liver and kidney damage and muscle damage around the heart. This was until quite recently the classic picture of Death Cap poisoning with over 90% of fungus-induced deaths being caused by this species. Now with new techniques of blood filtration by carbon-column haemodialysis units, and the use in America of the drug thioctic acid, effective treatment is possible if the poisoning is diagnosed in time. As mentioned previously *Amanita virosa* also has the same sort of toxins present and is just as deadly. There are a few other species which must be regarded as deadly such as *A. verna, A. bisporigera* and *A. tenuifolia*. Only *A. verna* is known to occur in Europe including Britain, the others being of North American origin. All are basically white species much like the Death Cap in shape. The distinguishing features of this group should be noted carefully: the basal bag-like volva and the ring on the stem, the white, free gills and the cap colour from olive-green through brown to pure white. In Europe the Death Cap is by far the commonest of these *Amanita* species and varies considerably in cap colour even within one woodland. In America the white species are more frequent, although the Death Cap does also occur.

Another genus recently discovered to contain amatoxins but not phallotoxins is *Galerina* (see pp. 110–11). Not all the species in the genus are poisonous – *G. mutabilis* illustrated in this book is edible and delicious – but other species, especially *G. unicolor* and *G. autumnalis*, have been implicated in poisonings in America (*G. unicolor* also occurs in Europe).

The muscarine poisons

A large number of species, including some other *Amanita* species, produce rapid, serious, but not usually fatal poisonings. Their toxins all affect the central nervous system to cause vomiting, profuse sweating, blurred vision, reduced heart beat and blood pressure and even convulsions with sometimes hallucinations as well! The best example of this type of mushroom must be the famous Fly Agaric, *Amanita muscaria* (pp. 52–3), used for centuries for its hallucinogenic properties and as a natural fly-killer. The toxins contained are muscarine, now thought not to be the active principal involved in the poisonings; plus muscimol and ibotenic acid. The toxic effect of this fungus varies enormously from region to region and with season, some forms being apparently harmless. However, the recent interest in hallucinogens should not encourage people to experiment, as the other side effects can be both unpleasant and dangerous. The Panther Cap, *Amanita pantherina* (also pp. 52–3), contains similar toxins. Muscarine itself although not present in large quantities in the amanitas is common in two other genera of gill-fungi, *Clitocybe* and *Inocybe* (see pp. 92–7 and 130–35). The poisonous species of these genera cause very serious poisonings, even death in some cases; *Clitocybe* includes a number of small, white, grassland species which contain this toxin (*C. rivulosa* is one illustrated), while *Inocybe* is almost entirely composed of poisonous species, mainly small brownish toadstools with fibrillose caps and brown gills.

For muscarine poisoning the specific antidote is atropine with other treatments for the various symptoms as they occur, while for muscimol and ibotenic acid symptomatic treatment is only with clorpromaxine for hallucinations, never atropine, which can aggravate an already dangerous situation.

Hallucinogenic toadstools

In Mexico Indians still eat certain toadstools to induce hallucinations, mainly species of *Psilocybe, Panaeolus* and *Conocybe* (pp. 80–81, 84–5, 158–9). These contain toxins named psilocybin which rapidly affect the senses, causing confusion, anxiety, hallucinations and often giddiness. Misidentification is very easy and could lead to serious poisonings.

Gastrointestinal irritants

A very large number of species cause digestive upsets some of which can be quite severe. However most of these irritant substances are destroyed by thorough cooking and some quite well-known edible species are in fact mildly poisonous if eaten raw or undercooked. The genera mainly responsible are *Russula, Lactarius, Tricholoma, Morchella* and even some boletes. Some true mushrooms (*Agaricus*) can cause upsets, even when cooked, in sensitive persons, especially *A. xanthodermus* (pp. 72–3), which is distinguished by its bright yellow stains when bruised. The author also has personal knowledge of sickness caused by eating *A. vaporarius* (pp. 72–3), although other people who ate it at the same table were unaffected.

Other toxins

The False Morel, *Gyromitra esculenta*, despite its specific name and being quite widely eaten without normally any ill effects, has caused severe poisonings, even death, in Europe. It contains a chemical called monomethylhydrazine (a constituent of rocket fuel!) which is usually evaporated during cooking. However if the fungi is undercooked it seems enough of the chemical remains to cause illness. It would also appear that ageing in this species sometimes causes an accumulation of the toxic principles so that old, perhaps slightly rotten, specimens are particularly dangerous.

Finally the common Ink Cap, *Coprinus atramentarius*, causes a strange poisoning only if consumed with alcoholic drinks. It contains a chemical recently named coprine which is similar in effect to the drug antabuse used in treating alcoholics. Symptoms include palpitations, rapid pulse, flushing of the extremities and face, nausea and vomiting. They soon subside only to recur when more alcohol is drunk, and this can last for several days.

All these cases sound alarming and indeed they are, but with common sense and caution, this account need not cause undue worry. Once again you *must know exactly* what you are eating to avoid risks. Should the worst happen (and small children particularly are liable to eat things when nobody is watching) seek medical aid immediately, and take any specimens of the offending fungus (even if only in the form of vomit) to the hospital with you for identification.

Collecting and Studying Fungi

A little time and preparation spent beforehand can often make a subject much more rewarding, and this is very true of mycology. Knowing where, how and when to collect makes a great difference to the success or failure of a day's study or quest for food. And when you have your day's collection safely home in good condition, how you go about handling the specimens, and the techniques used to investigate them, decide the chances of successful identification.

All woodlands of course will produce some fungi, and some will be very rich indeed, but you should pay particular attention when out collecting to the edges of woodlands and fields: these will have a much wider range of species than the centre and often produce rarities seldom seen elsewhere. The reasons for this are not always clear: temperature, moisture and plant variety no doubt all play a part. Similarly pathways and grassy rides passing through the woodland can also be very rewarding and make the day much more leisurely – scrambling through dense woods with thick undergrowth can be both unproductive of fungi and exhausting to the mycologist!

Mixed woodlands are better on average than pure stands of trees except where the latter are very ancient, mature woods or are on particular soils; beech on chalk, for example, can be extremely rich. The conifer plantations which cover extensive areas and which are relatively young can often be surprisingly poor except at the edges and very late in the season. Bogs and marshes have their own special flora, a very fascinating one to study although not usually producing the more showy species found in drier woodlands. Studies can very usefully be made of specialized habitats such as burnt stumps, sawdust heaps, sand dunes, dung, even back gardens! All such areas have a particular range of species determined by the chemistry of the soil or other growing medium, moisture content etc. and because of this it is often possible to predict exactly what species should be found by carefully noting the terrain. *Do not* collect more material than you can easily examine the same day; fungi are soft, easily decaying organisms and accurate identification depends on having perfect specimens. If you collect too many you will end up with only a portion identified and the rest too dried up or collapsed to be of use. Although collecting the fruit-bodies does not damage the fungus you should nevertheless show some consideration for others and not collect every specimen if a number are present; just take a representative selection of stages of growth and leave the rest for others to see.

A large flat basket or trug is ideal for collecting the larger toadstools; tins, tubes and boxes for very small and fragile species. Never use polythene bags which rapidly turn fungi

Spore shapes
a) fusiform, as in *Boletus* species; b) ornamented, as in *Russula* and *Lactarius* species; c) ovate; d) bullet-like, as in *Lepiota*; e) sausage-like, arcuate; f) globose, spiny; g) pip-shaped

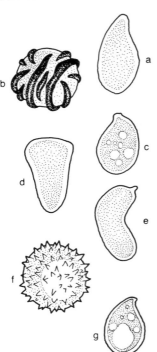

into a liquid mess; by far the best method is to wrap each toadstool individually in a twist of waxed paper of the kind found lining cereal boxes (not grease-proof paper). Have a selection of squares ready cut and wrap the fungi rather as a boiled sweet is wrapped in confectioners. This paper helps retain moisture, colour and stiffness in the fungus and also prevents the specimens from crushing each other in the basket. Always treat the fungus with great care when picking, particularly the stem which is easily damaged and is of great importance in identification. Make sure you uproot *all* the stem, especially the base, which can be vitally important if you intend to eat your finds: the stem is a key character in such deadly species as the Death Cap.

Take a written note in the field of any noticeable but often short-lived characters such as surface texture, odour and colour changes and look to see if the fungus appears to be growing under or by a particular tree or plant.

When you begin to examine your finds, try to develop the habit of examining them in a regular order, for example note the cap shape, texture, colour, presence of marginal veil, then gill spacing, attachment to stem, colour and so on all the way down. Finally when you have notes on all visual characters check the odour (which can often identify a puzzling specimen in seconds if very distinctive).

You should now take a spore print if spore colour is not obvious (gill colour can be a good guide but is *not always*; for example some species have purple gills and white spores). To take a print you just separate or cut the cap from the stem and place it, gills or tubes down, onto a piece of glass, cover with a tin or old cup to prevent loss of moisture and wait about half an hour to an hour after which enough spores should fall so that when scraped together (let dry for five minutes first) they will show their colour when placed over a sheet of pure white paper. These same spores can be examined under a microscope, if one is available, and their size measured. Never use spores from the gill surface for this purpose, unless a spore print is not available, as they may not be fully developed and so give you a false reading unless 40 or so are measured.

Always cut your specimens down the centre, which reveals the gill and tube attachments where relevant and also shows any flesh changes.

A drawing or painting of your finds, no matter how simple, will be a valuable source of reference in the future and is a good discipline in itself for remembering and noticing important features. Fungi can make beautiful subjects for anyone with artistic talent (whether this is already known to exist or not!).

Chemical tests are often a useful adjunct in identification and in some instances are species-specific, i.e. only one species in a genus will react to a particular chemical in a particular way. Useful reagents to keep are: Ferrous sulphate ($FeSO_4$) or ferric sulphate ($Fe_2(SO_4)_3$), both available as crystals (one large crystal in a tube is enough); ammonia (NH_4OH) in a 50% solution with distilled water; phenol, 2% in distilled water; formalin, 2% solution in distilled water; potassium hydroxide (KOH), 10% in distilled water; and a solution called Melzer's iodine, which is very important but will have to be prepared by a pharmacist with, in Britain at least, a doctor's written permission as it contains chloral hydrate, a dangerous poison. Its constituents are iodine 0.5 g, potassium iodide 1 g, distilled water 20 cc, mixed together, then chloral hydrate 20 g added. Some other less useful reagents are mentioned where necessary in the species descriptions. Many of these chemicals are dangerous, either poisonous or caustic or both. Handle them with great care, using a glass rod to apply, and store out of reach of children.

Finally, if you have easy access to a particular area of woodlands or fields try recording the species found in one restricted area, say a particular stand of trees or a stream-side, over a period of a number of years. You will often build up a surprisingly large list of species and only then get a true impression of the fungus flora of your area. It cannot be stressed enough that frequent and continual trips to an area are essential, if possible throughout the year, to reveal its full potential. Each season will be different and unique, revealing some new species or rarity, often in large numbers, sometimes not to be seen again for a great many years. This is the essence and the excitement of mycology in the field, not knowing what will be in the next field, the next woodland or roadside verge.

Classification

What groups of organisms should be included within the Kingdom Fungi, whether indeed there is a separate kingdom, and how the different groups relate to each other, are all matters which are undergoing intense re-examination. The following table and the classification used elsewhere in the book are broadly based on the view of Talbot (1971).

The Latin binomial name of a fungus, e.g. *Agaricus campestris*, consists of parts denoting the genus (*Agaricus*) and the species within it (*campestris*). In the text opposite the illustrations the better known of previously used names are given in parentheses following the sign =. Where the sign is lacking the name in parentheses is of subspecies status.

Within the kingdom (Mycota) the subsequent levels of classification (taxa) are, in ascending order of size: family (names ending in -aceae), order (-ales), class (-mycetes), subdivision (-mycotina) and division (-mycota). In the Guide section of the book, genus and family (sometimes order, not family) are indicated in the subheadings.

Myxomycota: slime moulds

Eumycota: "true" fungi

Mastigomycotina and **Zygomycotina**: "lower" fungi (formerly called "Phycomycetes"). Fungi with spores produced in asexual organs called sporangia. Includes water moulds, pin moulds, bread moulds, downy mildews etc.

Deuteromycotina: imperfect fungi or fungi imperfecti (includes some moulds). Fungi without sexual spores or with a sexual ("perfect") stage yet to be discovered; spores produced directly on the sterile mycelium or on specialized hyphae).

(All "larger" fungi are in the following two divisions)

Ascomycotina: "ascomycetes" or sac fungi. Fungi with sexual spores contained inside special cells called asci; usually forming fruit-bodies. Includes most of the fungal partners in lichens

Discomycetes: cup fungi
  Pezizales: cup fungi, morels, helvellas etc.
  Tuberales: truffles
  Helotiales: earthtongues and many plant parasites

Pyrenomycetes: flask fungi, including red bread moulds, apple canker, ergot
  Sphaeriales: includes candlesnuff fungus, cramp balls
  Hypocreales: includes coral spot

Plectomycetes: powdery mildews, black moulds, blue moulds; includes *Penicillium*
  Hemiascomycetes: yeasts etc.
  Loculoascomycetes: sooty moulds, plant scabs etc.

Basidiomycotina: "basidiomycetes". Fungi with sexual spores produced externally on cells called basidia; usually forming fruit-bodies

Holobasidiomycetes: fungi with simple (undivided) basidia
  Agaricales: agarics or gill fungi and allies, including most mushrooms and toadstools
  Boletales: boletes and allies
  Aphyllophorales: polypores, bracket fungi and allies, including chanterelles, fan, hedgehog, coral and fairy club fungi
  "Gasteromycetes": stomach fungi
  Phallales: stinkhorns
  Lycoperdales: puffballs and earthstars
  Sclerodermatales: earthballs
  Nidulariales: bird's nest fungi
  Hymenogastrales: false truffles

Phragmobasidiomycetes: fungi with septate (divided) basidia
  Tremellales: jelly fungi
  Auriculariales: including Jew's ear fungus
  Septobasidiales

Teliomycetes
  Uredinales: rusts
  Ustilaginales: smuts

How to Use the Key and Guide

In any book of this sort the choice of which species to include is a difficult one. Based as it is on the author's opinion as to which species are most likely to be found, and which species, although uncommon or rare, are worthy of inclusion, it must inevitably be a personal choice. Another mycologist might easily pick different species or disagree about which rare species should be included. The difficulty of choice arises because there are some 3,000 species of larger fungi in Britain alone. There will of course be a central core of species that are well known as common fungi everywhere and so must be included, but as more and more choices are made the decisions as to what must be dropped becomes an agonizing one. In this volume a conscious decision was made that where uncommon or rare species were involved we would try to show those not often illustrated in other similar works. Some of the species have in fact not been illustrated in a popular handbook before at all. Even some of the commoner species shown in the following pages are rarely seen or even mentioned in popular mycological works, purely because of the problem of limited space. It is hoped therefore that by the aid of this book you should be able to identify a very large proportion of the fungi you see and that your chances of identifying a rare species are increased. Where suitable, other related species to those illustrated have been mentioned in the text.

Assuming you have in front of you a fungus which you wish to identify, what is the best way to go about finding it in this book? Firstly it must be stressed that there is no reason at all why you should not just flip through the colour plates and try to spot your find. If you have a common or frequent species you stand a good chance of seeing it, but be sure to confirm the identification from the description opposite each plate.

Fungi are notoriously variable in appearance, and even a common species can vary remarkably. If difficulties or uncertainty arise, try tracking the fungus down in the Key that follows. The key is of the familiar dichotomous type, with paired choices or couplets of contrasting characters. To begin you should check the pictorial key to the different types of fungal shape and structure (pp. 34–5), and when satisfied as to which group a specimen belongs to, go to the appropriate section of the Key. So, if you have a simple disc-shaped fungus as shown in the pictorial key then follow on to step 7 in the written keys which deals with the cup fungi. For gill fungi you will need to make a spore print (p. 28), and also for some other species. The rules for using a key successfully are very simple: always read *both* contrasting couplets carefully and *only* follow a line which agrees entirely with your fungus. For example:

38. Cap viscid to glutinous, white . . . 39
38. Cap dry . . . 40

If your toadstool was white but *not* viscid-glutinous then you should follow the second choice. The number shown at the end indicates the next step, where you make your next choice. Should you find your fungus doesn't seem to come out correctly anywhere, there are various possible reasons why. Firstly, check very carefully that you haven't followed the wrong couplet or skipped a line (easily done); second, check (if you have more than one specimen available) that you are not using the oldest most faded specimen of those you have! A fresh, reasonably young, specimen is essential for a good chance of success. The third possibility is that you have a genus not included in the key. A key that includes all possibilities is both large and unwieldy and often requires detailed microscopical examination. With this key you should at least be able to discover to which genera it is close, and perhaps obtain assistance from a mycologist at a museum or botanical garden to complete the identification. This should not arise often as by far the majority of important genera are included. At no point does a couplet rely entirely on a microscopical character, although these have been included where useful. (The authorities used in spore descriptions are Lange and Hora, and Kühner and Romagnesi – see Bibliography.) A spore print to determine colour is however often essential.

Some of the genera included in the key have not been illustrated in this book, and are indicated as "(not incl.)". In other genera species may key out at different points as "(part)". This applies particularly to the Bracket fungi. Also where appropriate some fungi have been keyed out to species where distinctive or where only one species of the genus is known. The small marginal illustrations should help in placing key characters and shapes quickly and thus aid to quick and accurate identification. The newcomer to such keys should expect to make mistakes in the beginning, and try to double-check the identification with somebody else with more experience. Within a short time you should be able to recognize the principal genera by sight alone. Practice makes perfect!

Further Reading

General and illustrated works
Findlay, W. P. K., *Wayside and Woodland Fungi*, Frederick Warne, 1967. Good illustrations by Beatrix Potter and others, useful key to species.
Haas, H., *The Young Specialist Looks at Fungi*, Burke, 1969. Exceptionally good paintings of the commoner species.
Kibby, G., *The Love of Mushrooms and Toadstools*, Octopus, 1977. Good photos and descriptions of many common species.
Lange, M. and Hora, F. B., *Collins Guide to Mushrooms and Toadstools*, Collins, 1965. A standard illustrated guide for many years.
Ramsbottom, J., *Mushrooms and Toadstools*, Collins New

Naturalist, 1953. The most complete and scholarly as well as
thoroughly enjoyable book on all mycological subjects.
Still a must after many years. Not for identification.
Rinaldi, A. and Tyndalo, V., *Mushrooms and Other Fungi*,
Hamlyn, 1974. A comprehensively and beautifully illustrated
guide to over 1,000 species of Europe and North America.
A first-rate work.
Watling, R., *Identification of the Larger Fungi*, Hulton Group,
1973. A thorough and copiously illustrated guide to the
detailed study of fungi. No colour plates, but excellent black
and white microscopic studies.

Specialist works
Ainsworth, G. C., and Bisby, G. R., *A Dictionary of Fungi*,
Commonwealth Mycological Institute, 1971. An essential
reference work for the serious mycologist.
British Fungus Flora: Agarics and Boleti, various authors,
H.M.S.O., 1969– . Indispensable, with descriptions of all
British species of each group. Now available: by
R. Watling, P. Orton and D. M. Henderson, a general
introduction and key to families and genera; part 1, by
R. Watling, on Boletaceae, Paxillaceae, Gomphidiaceae.
Other parts in preparation.
Corner, E. J. H., *A Monograph of Clavaria and Allied Genera*,
O.U.P., 1950. A major work covering species of the world.
Dennis, R. W. G., *British Ascomycetes*, Cramer, 1968. A
standard work on the cup fungi of Britain and N. Europe
Dennis, R. W. G., Orton, P. D. and Hora, F. B., *New Check List
of British Agarics and Boleti*, parts 1 and 2, *Transactions of
the British Mycological Society*, 1960. An important reference
work. Both the *Bulletin* and the *Transactions* are worth
consulting for various articles, keys etc.
Kühner, R. and Romagnesi, H., *Flore analytique des
champignons supérieures*, Masson, 1953, republished 1974.
Still the most thorough and complete work of identification
for agarics, boletes, chanterelles; indispensable for the
serious collector but expensive and in French; black and
white line drawings, no colour.
Martin, G. W. and Alexopoulos, C. J., *The Myxomycetes*,
University of Iowa Press, 1969.
Pegler, D. N., *The Polypores, Bulletin of the British
Mycological Society*, vol. 7. Key to genera (and British
species).
Rayner, R. W., *Keys to the British Species of Russula, Bulletin
of the British Mycological Society*, vols. 2–4, 1968–70.
A thorough series for use both in the field and with a
microscope.
Revue de mycologie, Paris. This journal is another important
source of works on European fungi.
Talbot, P. H. B., *Principles of Fungal Taxonomy*, Macmillan,
1971.

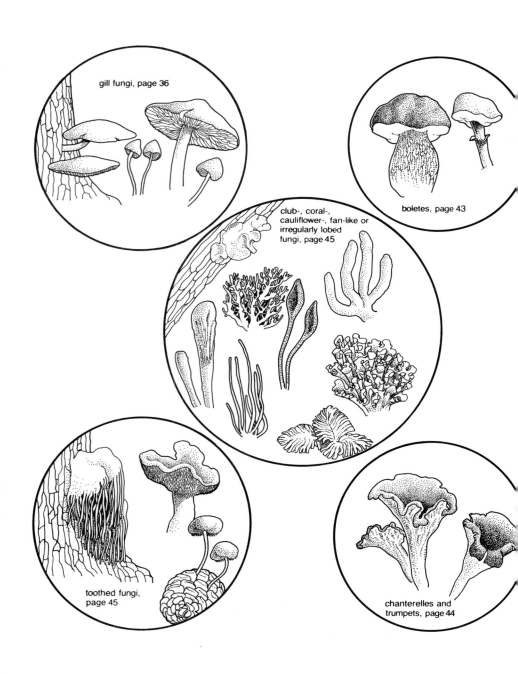

gill fungi, page 36

boletes, page 43

club-, coral-, cauliflower-, fan-like or irregularly lobed fungi, page 45

toothed fungi, page 45

chanterelles and trumpets, page 44

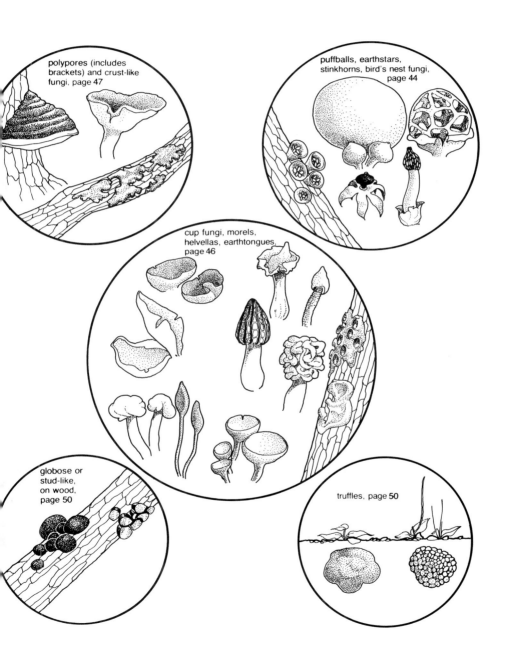

polypores (includes brackets) and crust-like fungi, page 47

puffballs, earthstars, stinkhorns, bird's nest fungi, page 44

cup fungi, morels, helvellas, earthtongues, page 46

globose or stud-like, on wood, page 50

truffles, page 50

GILL FUNGI

Spore colour (make a spore print, see p. 28)

a spores white, through shades of cream-colour, pale yellow to ochre, occasionally pinkish but never salmon-pink . . . 1

b spores from bright tawny orange through shades of brown (dull, rust, cigar, clay, ochraceous) to rich brown . . . 39

c spores black, sooty brown, chocolate or purplish . . . 59

d spores salmon-pink to lilaceous pink . . . 67

e spores red or blue-green . . . 73

lateral stem

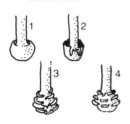

volvas

1 marginate   2 sac-like

3 hoop-like   4 wart-like

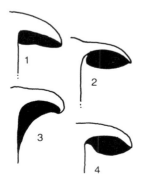

1 adnate   2 free

3 decurrent 4 sinuate

a spores white, through shades of cream-colour, pale yellow to ochre, occasionally pinkish but never salmon-pink

1 stem strongly eccentric to lateral, or absent . . . 34

1 stem central or at most only slightly eccentric . . . 2

2 with volva or volval remains at stem base and often on cap as warts, flakes etc.: *Amanita*

2 without a volva or volval remains . . . 3

3 ring present or cobweb-like or sheathing veil on stem. . . 4

3 without ring, or cobweb-like or sheathing veil on stem . . . 13

4 granular veil sheathing lower stem up to slight ring-like zone (latter may be absent); gills adnate to sinuate: *Cystoderma* (part), *Lepiota* (part)

4 veil, if sheathing, not granular *or* with ring only and gill attachment may be different . . . 5

5 veil sheathing and usually slimy; gills rather thick, waxy; all species on ground: *Hygrophorus* (part)

5 with membranous ring or cobweb-like veil . . . 6

6 gills adnate to slightly decurrent; cap and stem finely squamulose to scaly; medium to large species usually tufted on wood, attached by black bootlace-like rhizomorphs which spread under wood and through soil: *Armillaria* (part)

6 gills not decurrent and/or other characters different . . . 7

7 gills free . . . 8

7 gills adnate-adnexed to sinuate . . . 10

8 cap smooth, viscid; ring membranous; on ground: *Limacella* (not incl.)

8 not this combination of characters; cap usually dry, mostly with scales/fibrils, especially at centre; on ground . . . 9

9 delicate, tiny to very large fleshy species, with ring-zone to thick double collar-like ring: *Lepiota* (part), *Cystoderma* (part, not incl.)

9 delicate, graceful, very fragile species; cap noticeably striate-sulcate at margin: *Leucocoprinus* (not incl.)

10 cap viscid to glutinous, white, ivory to greyish . . . 11

10 cap dry or at most moist (if gills thick, waxy, growing on soil, see *Hygrophorus*) . . . 12

Leucocortinarius bulbiger

Russula fruit-body and spore

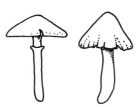

typical Hygrophorus   typical Hygrocybe

Clitocybe

11 on beech; cap white, ivory to greyish, glutinous; gills broad, widely spaced; stem slender, very tough with distinct narrow ring: *Oudemansiella mucida*

11 on ground; cap cream-colour to tan, slightly viscid; gills not widely spaced; stem with faint ring-zone and lower stem often "weeping" rust-coloured droplets, especially in damp weather: *Chamaemyces fraccida* (not incl.)

12 stem with large marginate basal bulb; cortina leaving ring-zone; cap reddish brown; on ground; like a white *Cortinarius: Leucocortinarius bulbiger* (not incl.)

12 stem without marginate bulb; with fine woolly ring at apex; cap grey, fibrillose; under willows: *Tricholoma cingulatum*

13 cap, stem and gills brittle, crumbly, rather granular, *not* fibrous; spores ornamented with spines and ridges which stain black in iodine . . . 14

13 cap, stem and gills soft to touch but fleshy and fibrous, *not* crumbly and granular; spores different . . . 15

14 broken flesh exudes white or coloured "milk"; taste often distinctive (peppery, sweet etc.): *Lactarius*

14 flesh not exuding milk when cut; cap colours widely variable – bright red, purple, yellow, green, brown, white etc.: *Russula*

15 gills thick, rather distant, pinkish, reddish brown or violet, dusted white with spores; on ground in woods, heaths etc.; small to medium species; spores minutely spiny: *Laccaria*

15 gills (if thick) of different colour, not noticeably dusted with spores, often waxy or translucent in appearance *or* not particularly thick and distant . . . 16

16 gills rather thick with texture waxy to watery and translucent appearance: *Hygrophorus* (part), *Hygrocybe*

16 gills not particularly thick and/or not waxy and translucent . . . 17

17 gills adnate to decurrent (if clustered on wood, see *Xeromphalina* (not incl.), *Armillaria tabescens*, *Mycena* species) . . . 18

17 gills free, adnexed-adnate or sinuate . . . 22

18 gills forking, pale to deep orange, edge blunt and rounded; cap colours similar; in heaths under birch, bracken etc.: *Hygrophoropsis aurantiaca*

18 colour different *or* gills not forking and blunt . . . 19

19 gills usually blunt, spotted reddish; cap grey to blackish; spores blue-black in iodine: *Cantharellula umbonata* (not incl.)

19 gills not usually forking although one or two may do so, edge sharp and well-formed . . . 20

20 cap small (2–3 cm) to very large, often pale whitish to grey-brown; stem fibrous, not "polished", over 5 mm in diameter: *Clitocybe, Leucopaxillus*

20 usually very small species, cap below 2–3 cm; stem smooth, "polished", under 5 mm diameter . . . 21

21 usually in moss or on bare soil; cap often rounded with central depression to funnel-shaped; spores not blue-black in iodine; gills rather widely spaced; stem usually short: *Omphalina*

21 on pine stumps or litter in clusters; small tough species, cap convex; spores blue-black in iodine: *Xeromphalina* (not incl.)

22 flesh tough and leathery, can dry out then revive with water without decaying; gills often interconnected with "veins"; very tiny to medium-sized species: *Marasmius* (and some related genera, all small to medium, not incl.)

22 flesh, although it may be tough, will *not* so revive . . . 23

23 cap tiny to medium-sized, bell-shaped to umbonate, often sharply so, not fleshy; margin usually noticeably striate, not incurved when young; gill attachment variable, adnate to slightly decurrent; colours variable – browns, greys, yellow, pink, white all frequent; often tufted, but solitary species also; on wood and ground: *Mycena*

23 characters different . . . 24

Mycena

24 stem long, tough, fibrous, with extended "taproot" below soil; cap viscid-glutinous *or* velvety-hairy, with radial wrinkles; usually by stumps: *Oudemansiella* (part)

24 stem without "taproot" *or* other characters different . . . 25

25 cap deep reddish brown; stem swollen, spindle-shaped, *very* tough, fibrous, splitting, misshapen; often in clusters at base of deciduous trees: *Collybia fusipes*

25 colour different and/or stem different from preceding . . . 26

26 clustered or solitary on wood . . . 27

26 on ground or other fungi . . . 30

27 in dense clumps on deciduous trees; caps 4–10 cm, minutely scaly at centre; stems slender; overall colours tan to brownish: *Armillaria tabescens*

27 characters very different *or* on coniferous timber . . . 28

Oudemansiella

28 on deciduous timber; caps rather small (2–6 cm), yellow-orange, smooth and rather viscid; stem slender, yellow above, reddish brown, velvety below, black at base; late autumn through winter: *Flammulina velutipes*

28 on deciduous or coniferous timber, also sawdust; medium to large fleshy species; other characters different . . . 29

29 on conifers and sawdust; cap yellow overlaid with minute purple-red to brownish downy scales; gills golden yellow: *Tricholomopsis* (part)

29 on deciduous timber; caps greyish brown, smooth but radially fibrillose, not scaly; gills whitish, broad, distant; stem white, tough and fibrous, often with white thick fibrous "roots" or runners: *Tricholomopsis platyphylla*

30 in dense clusters with stem-bases often fused; caps white to grey-brown; medium to large and fleshy, texture elastic-rubbery; stem fibrous; gills adnate to very slightly decurrent: *Lyophyllum* (part)

Lyophyllum

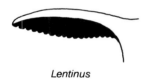

Lentinus

30 not in such dense clusters with fused basal mass, although may occur in tufts with bases adhering or smaller, more slender species . . . 31
31 growing on other, old fungi: *Nyctalis*
31 not growing on other fungi . . . 32
32 gills sinuate or sinuate-adnate; cap often radially fibrillose to scaly, occasionally smooth: *Tricholoma, Melanoleuca, Calocybe*
32 gills adnexed, adnate or almost free; cap usually smooth, never fibrillose or scaly . . . 33
33 strong odour of new meal, flour, cucumber, or rancid; entirely greyish, grey-brown to black; rather tiny to small species: *Lyophyllum* (part, not incl.)
33 without these odours; colours usually white through shades to brown; medium to large species: *Collybia* (part)
34 edge of gills serrated or split lengthwise . . . 35
34 edge of gills entire (not broken) . . . 36
35 edge of gills serrated, notched: *Lentinus, Lentinellus*
35 edge of gills split lengthwise; without a stem; simple bracket-like cap: *Schizophyllum commune* (not incl.)
36 fruit-body rather tongue-shaped with short lateral stem; spores blue-black in iodine: *Panellus*
36 fruit-body shape different *or* longer stem, spores not blue-black in iodine . . . 37
37 gills yellowish to orange; cap orange-yellow, tomentose: *Phyllotopsis nidulans* (not incl.)
37 not this combination of characters; caps often clustered or overlapping, sometimes fused; stem variable in length, often absent . . . 38
38 very tough, woody when dry; irregularly shaped caps clustered and often fused; on deciduous stumps; brown with violet down on stem when young: *Panus torulosus*
38 never hard and woody; caps distinct, bracket-like to rounded, even funnel-shaped, with or without stem of variable length, eccentric to lateral; on living or felled trees; colours usually white through grey to brownish or blue-grey: *Pleurotus*

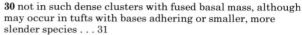

b spores from bright tawny orange through shades of brown (dull, rust, cigar, clay, ochraceous) to rich brown
39 cap bracket-like, with or without tiny lateral stem: *Crepidotus, Paxillus* (part)
39 stem central to slightly eccentric, not lateral . . . 40
40 with distinct membranous ring on stem . . . 41
40 ring fine, cobwebby, often rapidly vanishing, or entirely absent . . . 45
41 very large species 15–25 cm high; yellow-orange colours; cap texture granular; on soil: *Phaeolepiota*
41 not this combination of characters . . . 42

Panus torulosus

Hebeloma radicosum

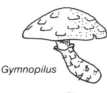

Gymnopilus

Pholiota

Inocybe

Paxillus

42 stem deeply rooting; associated with animal or bird burrows in ground near tree roots; cap pale ivory to tan with irregular vague filmy scales; strong bitter-almond (marzipan) odour when fresh: *Hebeloma radicosum*
42 with different characters from preceding . . . 43
43 stem not scaly; cap white to dull brown, smooth or wrinkled, not scaly or scurfy; spores dull cigar- or chocolate-brown: *Agrocybe* (see also *Rozites*)
43 stem often scaly; cap mostly yellowish to orange, occasionally whitish; spores tawny orange to rust- or ochre-brown . . . 44
44 spores tawny orange to rust; only on wood (sometimes buried) at base of trees, never high up; gills bright yellow-orange, often spotted rust-brown; stem stout, often bulbous: *Gymnopilus junonius*
44 spores ochre- to rust-brown; usually on wood (often high up on trees), occasionally soil; gills not usually with rust-like spots; stem frequently scaly; cap dry to glutinous: *Pholiota* (part) (see also *Rozites*; *Galerina* species tufted on wood, i.e. *G. mutabilis*, pp. 110–11, may key out here, but cap is hygrophanous)
45 distinct cobweb-like veil joining cap-margin to stem when young, often left as fine cobwebby ring or zone . . . 46
45 without cobweb-like veil *or* if it is present, it is very faint and soon lost, not leaving a ring zone . . . 49
46 on wood, often tufted; spores tawny orange to ochre-brown; cap shades of orange, tan to orange-brown, minutely scaly and dry to smooth, sometime glutinous: *Gymnopilus* (part), *Pholiota* (part)
46 not on wood; spores ochre-, rust- to dull cigar-brown . . . 47
47 spores rust-brown; often with copious cobweb-like veil; cap almost any colour (any fungus with rust-brown, not ochre, spores, and bluish colours is a member), texture dry to glutinous or hygrophanous: *Cortinarius* (some *Pholiota* and *Rozites* species may be confused)
47 spores ochre- to dull cigar-brown . . . 48
48 usually cap radially fibrillose to rough and minutely scaly, mostly dry, and stem fibrous; odour often earthy, unpleasant, to strongly fruity (e.g. of pears); spores cigar- to dull brown, strongly nodulose, or smooth and elongate to bean-shaped: *Inocybe*
48 cap mostly smooth, or only fibrils of veil at margin, whitish to clay or brick-red, often viscid; frequent odour of radish, or sweetly sickly; spores clay- to ochre-brown, ovate-elliptic: *Hebeloma* (part)
49 gills decurrent, soft, bruising brown; medium to large brown cap, margin inrolled, tomentose: *Paxillus* (part)
49 gills not decurrent *or* if so, then cap whitish . . . 50
50 gills free; very delicate, fragile species; cap bell-shaped to flat, strongly striate, egg-yellow, soon appearing to dissolve away; on dung, rich soil, grass: *Bolbitius vitellinus*

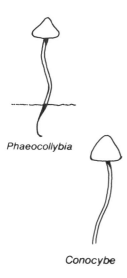

Phaeocollybia

Conocybe

50 gills not free and/or other characters different . . . 51
51 cap conical to bell-shaped . . . 52
51 cap more or less broadly convex to flattened . . . 55
52 cap sharply pointed, conical; stem rooting: *Phaeocollybia* (not incl.)
52 cap bluntly conical; stem not deeply rooted . . . 53
53 odour strong of fish or cucumber; cap brownish; stem dark, slender: *Macrocystidia cucumis*
53 without odour of fish or cucumber . . . 54
54 cap surface dull or shiny, cap rather deep, campanulate or almost cylindric; stem tall (5–8 cm), very slender, straight; in grass; cap-surface cells rounded: *Conocybe*
54 cap more or less shiny, even viscid; stem short, slender, often curving; usually in moss; cap-surface cells filamentous: *Galerina*
55 cap roughened with minute pointed upright scales: *Phaeomarasmius* (not incl.)
55 cap smooth or with whitish scales at margin only . . . 56
56 cap whitish; gills decurrent; cap margin slightly hairy; spores clay-brown: *Ripartites* (not incl.)
56 cap whitish to brown; gills not or only slightly decurrent . . . 57
57 cap white to clay-brown or brick, more or less smooth, margin inrolled at first, often with fibrils from veil; gills and spores finally dull clay- or ochre-brown; often with radishy odour: *Hebeloma* (part)
57 not this combination of characters; cap pale to rich brown, with or without vague scurfy marginal scales . . . 58
58 cap pale whitish brown to ochre with faint marginal scales; small species on bare soil; spores smooth, thin-walled; gills may be slightly decurrent: *Tubaria* (not incl.)
58 cap ochre to brown, not scaly; under alders or willows; spores minutely warted, rough: *Naucoria* (not incl.)

Coprinus

Agaricus

c Spores black, sooty-brown, chocolate or purplish
59 gills and cap dissolve into inky liquid on maturity (*or* tiny grooved caps in very large clusters); tiny to very large: *Coprinus*
59 gills not dissolving into inky liquid . . . 60
60 gills thick, broadly spaced, decurrent; cap viscid to glutinous; medium to large: *Chroogomphus, Gomphidius*
60 gills not thick, decurrent . . . 61
61 gills mottled black on grey because of uneven spore ripening; stem slender, stiff, straight, easily snapping; cap more or less convex to bell-shaped: *Panaeolus*
61 not this combination of characters . . . 62
62 gills free, often remote from stem; on ground, never wood; usually with membranous ring or at least a ring zone; cap white to brown, smooth to scaly; small to very large: *Agaricus* (true mushrooms)
62 gills adnate, adnexed or sinuate; on ground or wood . . . 63

63 gills mostly adnate, purple-brown; stem usually with ring or veil; cap blue-green, yellowish to brown or reddish, often viscid-glutinous or dry with marginal scales; on dung, soil, woodchips or in grass: *Stropharia*

63 gills not adnate and/or other characters different . . . 64

64 stem and cap-margin with abundant cobweb-like veil; gills black, edge often "weeping" black droplets in damp weather; cap ochre to orange, woolly-fibrillose; on ground: *Lacrymaria*

64 not the above combination of characters . . . 65

65 cobweb-like veil when young but not a membranous ring; gills sinuate to adnate; often tufted on wood, medium-sized, fleshy *or* solitary and slender on ground; cap dry, more or less smooth, brightly coloured yellow or reddish to brown; spores purple-brown: *Hypholoma*

65 gills usually adnate; generally fragile species, thin-fleshed; on wood or ground . . . 66

66 very fragile with delicate cap conical to convex, then flattened, usually greyish to brown, hygrophanous; gills adnate; clustered or solitary, on ground or wood; veil present or not, may form ring and/or scales on stem and cap: *Psathyrella*

66 cap convex to sharply pointed, margin incurved when young; colours yellowish to brown; gills adnate, often almost triangular; stem stiff, slender, breaking with a snap: *Psilocybe*

Psilocybe semilanceata

d spores pale to deep salmon-pink

67 stem to one side of bracket-like cap: *Claudopus* (angular spores) and *Rhodotus* (pink to apricot wrinkled cap; spores warted) (neither incl.)

67 stem central to only slightly eccentric . . . 68

68 stem with volva (thin, easily lost when picked): *Volvariella*

68 without cup or volva . . . 69

69 gills deeply decurrent: *Eccilia* (not incl.), *Clitopilus*

69 gills free *or* if attached not truly decurrent . . . 70

70 gills free; cap and stem easily separable; on wood or soil: *Pluteus*

70 gills attached to stem for part or all of their depth . . . 71

71 gills sinuate to adnate; medium to large fleshy species; cap convex-flattened, not pointed; usually late autumn–early winter; spores ovate, prickly: *Lepista*

71 usually small to medium, fibrous and rather fragile, generally not thick fleshy species; all with angular many-sided spores; gills mostly adnate-adnexed, or sinuate . . . 72

72 gills more or less sinuate; cap medium-sized, usually pointed, or with rather raised centre, fibrillose: *Entoloma*

72 small, fragile caps, rounded or bell-shaped to pointed; gills adnexed-adnate: *Nolanea* (smooth caps, not incl.), *Leptonia* (scaly cap and/or bluish colours)

Volvariella

Entoloma spore

e spores red or blue-green
██ **73** spores and gills blue-green; delicate, small, with granular veil: *Melanophyllum eyrei* (not incl.)
██ **73** spores and gills reddish; small stocky species; stem with granular veil: *Melanophyllum echinatum*

BOLETES

The few Boletales with gills are included in Gill Fungi, above
1 thick overlapping scales on cap; stem rough, shaggy; whole fungus grey-black; flesh reddens when cut: *Strobilomyces floccopus*
1 cap without scales *or* scales not grey-black . . . 2
2 cap with pointed umbo, covered with fine pointed scales (scabrosities), yellow-ochre to rust-brown; pores large, honeycomb-like, yellow; stem with ring at top; under larches: *Boletinus cavipes*
2 not this combination of characters . . . 3
3 stem covered with small woolly tufts which are usually white, then reddish brown to black: *Leccinum*
3 stem smooth or sticky, with fine network, sometimes ring at top . . . 4

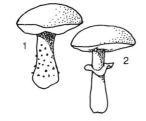

1   2

1 *Leccinum*   **2 *Suillus***

4 cap usually sticky to glutinous; stem with or without ring; always connected to various conifers: *Suillus*
4 cap at most moist to viscid in damp weather only; never with ring; usually below deciduous trees . . . 5
5 mature pores pink (not red, buff) as is spore-mass; stem with raised network; taste bitter: *Tylopilus felleus*
5 pores not pink, usually some shade of yellow, white, greenish or orange to red; often bruising blue . . . 6
6 cap and stem smooth, greyish black; spores purple-brown: *Porphyrellus pseudoscaber*
6 not blackish; spores yellow to ochre or brown . . . 7
7 pores vivid chrome to golden, not bruising blue; cap moist, clay- to strawberry-pink, 2–5 cm: *Aureoboletus cramesinus*
7 pores duller yellow or quite different colour, often bruising blue; cap not pinkish, mostly over 5 cm across . . . 8

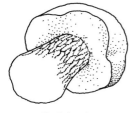

reticulate stem

8 stem usually hollow; flesh firm, brittle; pores white to pale yellowish; texture of cap skin velvety or roughened; spores pale yellow: *Gyroporus*
8 stem solid, never clearly hollow; other characters mostly different . . . 9
9 always attached to earthballs (*Scleroderma*, pp. 236–7); small, yellow-brown: *Boletus parasiticus*
9 not on *Scleroderma* or other fungi; stem smooth or reticulate; pores white to yellow or red; cap whitish, clay, brown, to reddish; flesh and pores often bruising: *Boletus*

Gyroporus

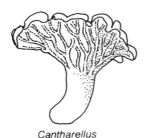

Cantharellus

CHANTERELLES AND TRUMPETS

Smooth or wrinkled undersurface, no true sharp-edged gills
1 grey-black, trumpet-like; almost smooth on outer surface,
hardly any wrinkles or folds: *Craterellus cornucopoides*
1 distinct folds, wrinkles or blunt very irregular "gills" below
flared, irregularly shaped cap; colours various, grey to brown
or yellow-orange . . . **2**
2 small (2–5 cm across), dark brown; grey, wrinkled lower
surface: *Cantharellus cinereus*
2 almost gill-like blunt wrinkles and folds; cap brown, orange
or apricot, occasionally flushed violet . . . **3**
3 cap brown; stem yellow to orange: *Cantharellus
infundibuliformis*
3 entirely apricot-orange; thick-fleshed: *Cantharellus cibarius*
Some less common species (not incl.) occur, differing mainly
in colour (e.g. *C. amethysteus* with violet flush and scales)

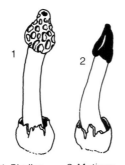

1 *Phallus*   2 *Mutinus*

PUFFBALLS, EARTHSTARS, STINKHORNS, BIRD'S NEST FUNGI

Fungi without gills or pores; spores in powdery central mass
or as smelly slime or as "eggs" in tiny "nest"
1 mature fungus has slimy, foul-smelling spore-mass spread
over cap or cage-like arms; if it starts as a whitish "egg", then
with jelly-like layer under skin . . . **2**
1 mature fungus without foul odour; mature spores usually
powdery or in tiny egg-like masses . . . **4**
2 fruit-body breaks out from "egg" (size of hen's); stem bears
separate thimble-shaped cap covered at first with green-black
slime: *Dictyophora* (lace-like veil hanging from cap), *Phallus*
2 "egg" and fruit-body much smaller *or*, if bigger, fruit-body
reddish, cage-like . . . **3**
3 starts as a small (1–3 cm) "egg", sends up spongy pointed
stem without separate cap: *Mutinus caninus*
3 arises from large (5–8 cm) "egg"; fruit-body cage-like,
reddish, with smelly greenish slime on inside: *Clathrus ruber*
4 fruit-body like a brownish onion, splitting open to form a
star shape with a rounded puffball at the centre: *Geastrum*
4 fruit-body globose, with stem or not *or* like tiny "nest" with
"eggs" . . . **5**
5 very large (10–30 cm) globose fruit-body; skin smooth, white,
leathery, cracking with age; in fields and hedgerows:
Langermannia gigantea
5 smaller, if rounded, *or* different shape entirely . . . **6**
6 tiny (2 cm or less) "bird's nest" with one to ten "eggs" . . . **10**
6 rounded, 2 cm or more; with or without a base or stem . . . **7**

Geastrum

Scleroderma

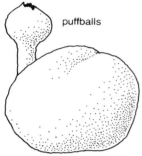

puffballs

bird's-nests

Auriscalpium

Hydnum

7 on soil; round; skin thick, tough, usually scaly or warty, sometimes smooth, cream to yellow-ochre or brownish; mature spore-mass inside purplish to black or olive-brown with strong but not foul odour: *Scleroderma*

7 on soil or grass, occasionally wood; skin much thinner; tiny warts, spines or scales, if present, are easily brushed off or broken; spores puff out when mature (puffballs) . . . 8

8 fruit-body round or with basal stem; usually whitish, often with fragile warts or spines; spore-dispersal through apical pore: *Lycoperdon, Bovista*

8 with thin tough stem buried in sandy soil *or* without single apical pore (skin flakes with age on upper parts) . . . 9

9 puffball (1–3 cm) just projecting above sandy soil with thin buried stem below: *Tulostoma*

9 puffball (3 cm or more) with stem or not; upper part flaking away to release spores, often just base remaining: *Calvatia*

10 tiny star-shaped cup or "nest" 1–3 mm across with one tiny "egg": *Sphaerobolus* (not incl.)

10 "nest" rounded, larger (1–2 cm), with some 5–10 "eggs" . . . 11

11 "nest" deep brown to grey, funnel-shaped, flaring: *Cyathus*

11 "nest" straw-yellow to ochre, cup-shaped: *Crucibulum*

TOOTHED FUNGI

Pendent teeth or spines on lower surface

1 on trees, logs etc . . . 2

1 on ground or pine cones . . . 3

2 large, irregularly branched; pure white to ivory with long to very long pendent spines; on usually dead deciduous and coniferous trunks: *Hericium* and other rarer genera

2 tongue-shaped, soft, jelly-like; whitish grey with short soft teeth; on pine stumps: *Pseudohydnum gelatinosum*

3 on soil in leaf-litter or pine needles; medium to large; stem thick . . . 4

3 on pine cones; very small, spoon-shaped; stem slender, lateral: *Auriscalpium vulgare*

4 cap large with thick coarse scales, deep grey-brown: *Sarcodon imbricatum*

4 cap small to medium, reddish brown to pinkish, smooth or at most cracked, never distinctly scaly: *Hydnum* and related genera

Many rarer genera of terrestrial toothed fungi are to be found

CLUB-, CORAL-, CAULIFLOWER- OR FAN-LIKE OR IRREGULARLY LOBED

1 simple unbranched club, sometimes clustered . . . 2

1 branched clubs or cauliflower-, coral- or fan-shaped . . . 6

2 on the ground . . . 3

2 on deciduous wood . . . 4

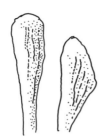

Clavariadelphus

1 *Xylosphaera*   2 *X. polyporpha*
hypoxylon

3 club tall (10–25 cm), swollen, sometimes flattened above; usually in beech litter: *Clavariadelphus*
3 club shorter, more slender, blunt and club-like to sharply pointed *Clavaria, Clavulina* (part, not incl.), *Cordyceps* (not incl.), *Clavulinopsis* (part)
4 tiny bright yellow-orange pointed clubs: *Calocera cornea*
4 colour different . . . 5
5 small (1–5 cm high) pointed and flattened clubs (sometimes branching, antler-like), white, base blackened, tips like burnt candle wick: *Xylosphaera hypoxylon*
5 larger (3–8 cm) clubs, clustered; hard, black, swollen, often finger-like: *Xylosphaera polymorpha*
6 branched, antler-, fan- or coral-like . . . 7
6 cauliflower-like or with soft irregular lobes . . . 9
7 on bare soil or pine needles; dull soft brown; irregularly branched and flattened, fan-like: *Thelephora*
7 on soil, coral-like *or* on wood, colour and shape different . . . 8
8 on soil (or on wood, olive-colour, never yellow-orange); coral-like, often brightly coloured: *Ramaria* (spores ornamented), *Clavulina* (part), *Clavulinopsis* (part, not incl.)
8 on wood; in loose or dense clusters of slender, pointed and branched clubs: *Calocera viscosa* (bright yellow-orange), *Xylosphaera hypoxylon* (white and blackish)
9 large, rounded, cauliflower-like, with many crisp flattened lobes, pale tan; on or by conifers, rarely other trees: *Sparassis crispa* (see also *Peziza proteana*)
9 soft, gelatinous, much more irregularly lobed, 5–10 cm across; yellow or deep brown; on deciduous or coniferous timber: *Tremella*

Thelephora

Cyathipodia macropus

CUP FUNGI, MORELS, HELVELLAS, EARTHTONGUES

1 with a stem, long or short, and a head or cap . . . 2
1 without a stem; a simple cup-, disc- or top-shape . . . 10
2 simple cup 2–6 cm across on definite stem either narrow or stout: *Cyathipodia macropus*, some small *Peziza* (see 16 below)
2 not a simple cup; head usually club- or tongue-like or much more complex . . . 3
3 head simple, tongue-shaped and flattened *or* rounded or button-shaped, gelatinous . . . 4
3 cap complex, wrinkled, sponge-, saddle- or thimble-shaped . . . 6
4 head tongue-shaped, flattened, *not* separated from stem; green to blackish: *Microglossum, Trichoglossum* or (similar with rounded head on insect larva) *Cordyceps*
4 head a soft irregular gelatinous "button" *or* tongue-like but clearly divided from stem . . . 5

Leotia

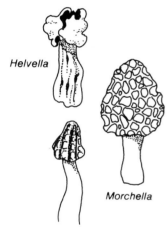

Helvella

Morchella

Mitrophora

5 cap rounded, button-like, margin inrolled, olive-green; yellowish central stem . . . *Leotia* (in deciduous woods) *or* the same but yellow with grey stem and in coniferous woods . . . *Cudonia*

5 bright yellow flattened or tongue-shaped head with paler stem pushing up into one edge; in conifer woods . . . *Spathularia or* the same but orange head, by ponds, swamps . . . *Mitrula*

6 cap wrinkled, brain-like *or* convoluted, saddle-shaped . . . 7

6 cap like a sponge with pits and ridges *or* smooth, thimble-shaped . . . 8

7 cap wrinkled, brain-like: *Gyromitra*

7 cap convoluted, often saddle-shaped: *Helvella* (part)

8 cap smooth, thimble-shaped: *Verpa*

8 cap like a sponge with pits and ridges . . . 9

9 cap base fused with stem: *Morchella*

9 cap base not fused with stem: *Mitrophora*

10 lop-sided, rabbit's ear-like cup: *Otidea*

10 rounded or irregular cup, or disc- to top-shaped, but not like a rabbit's ear . . . 11

11 tiny scarlet disc 0.5 cm across with fringe of black lashes or hairs at edge: *Scutellinia*

11 not with the above combination of characters . . . 12

12 very irregular gelatinous mass forming small clusters of top-shaped fruit-bodies 0.5–1 cm across, deep violet; shapeless when immature; on wood: *Ascocoryne sarcoides*

12 cup or disc, often irregular; on soil or wood . . . 13

13 a velvety brown to purplish "ear" on wood; elastic and rubbery: *Hirneola auricula-judae*

13 shape and texture (fragile, brittle) different . . . 14

14 tiny (0.5 cm) bright green cups; usually on oak, staining wood green: *Chlorociboria aeruginascens*

14 not with the above combination of characters . . . 15

15 deep brownish black, disc-shaped, then edges turned downwards, cushion-shaped; very short white root-like structures below; on burnt sites under conifers: *Rhizina*

15 without this combination of characters . . . 16

16 bright orange medium to large cups; often by pathsides: *Aleuria aurantia*

16 small to large cups and discs, coloured cream, tan, brown, orange-red, violet etc.; on soil or wood and other rotting matter: *Peziza, Sarcoscypha* and other genera of smaller cup fungi

POLYPORES (includes brackets) AND RESUPINATE (crust-like) FUNGI

On wood or soil; tough, fleshy or woody; with pores, or just wrinkled surfaces clinging to wood

1 with a distinct stem, either central or lateral . . . 31

1 without a distinct stem . . . 2

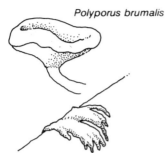

Polyporus brumalis

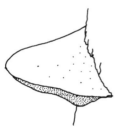

resupinate/crust-like

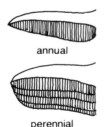

bracket

annual

perennial

2 resupinate to bracket-shaped; distinct pore layer . . . 9

2 bracket-like or more often entirely resupinate or resupinate with edge protruding to form thin poorly shaped bracket; without distinct pores . . . 3

3 resupinate, often forming irregular brackets; lower surface wrinkled, without shallow "pores"; firm, fleshy . . . 5

3 bracket-like or (usually) resupinate with ill-formed brackets; lower spore-producing surface wrinkled and vein-like, forming shallow "pores"; soft, elastic or rubbery . . . 4

4 spores white: *Merulius*

4 spores brown: *Serpula* (not incl.) (contains Dry Rot)

5 lower surface of fruit-body (irregular brackets) contains long brown spines: *Hymenochaete* (not incl.)

5 without spines . . . 6

6 outer surface of the resupinate fruit-body usually radially veined, often bright orange: *Phlebia*

6 outer surface not radially veined . . . 7

7 spores brown; fruit-body resupinate, brownish and rather lumpy at centre, yellowish white at margin: *Coniophora* (not incl.) (contains Wet Rot or Cellar Fungus)

7 spores white . . . 8

8 fleshy; either resupinate or with brackets: *Stereum, Chondrostereum* and some related genera

8 not fleshy; without distinct brackets, often a very thin crust or even coloured "wash" on wood; various corticiaceous fungi (not incl.)

9 tubes free from each other, easily separated; fruit-body a tongue-shaped fleshy red bracket; flesh red, often "bleeds" when cut: *Fistulina hepatica*

9 tubes fused to form a united pore layer . . . 10

10 perennial brackets, building up more than one layer of tubes; mostly hard, woody . . . 11

10 annual brackets (occasionally resupinate), forming one tube layer only before drying or decaying; usually fleshy . . . 14

11 spores brown . . . 12

11 spores white . . . 13

12 large extremely hard brackets on deciduous timber; chesnut brown, margin whitish; pores white, bruise brown: *Ganoderma* (part)

12 woody brackets on conifers (usually several feet up); deeply cracked upper surface: *Cryptoderma* (not incl.)

13 large (12–25 cm) thich, very hard woody brackets, hard upper crust: *Fomes*

13 small to medium (5–12 cm) brackets, generally fleshy to corky not woody, with upper crust or surface slightly furry: *Oxyporus, Fomitopsis, Heterobasidion* (none incl.)

14 pores very elongate to labyrinth-like *or* bracket not regularly formed, mostly resupinate instead . . . 15

14 pores distinct and rounded, or only slightly elongate; in regular brackets . . . 19

15 pores labyrinth-like: *Daedalea, Daedaleopsis*
15 pores very elongate, almost like thick woody gills, *or* fruit-body resupinate . . . 16
16 elongate, gill-like pores . . . 18
16 without bracket, or else very irregular caps and mainly a resupinate pore layer . . . 17
17 completely resupinate; not very fleshy: *Fibuloporia* and related genera (none incl.)
17 resupinate but with irregular caps at margin; fleshy and tough: *Datronia, Gloeoporus, Bjerkandera* (part) (none incl.)
18 bracket whitish to pale brown: *Lenzites*
18 bracket brown: *Gloeophyllum* (not incl.)
19 pores bright red, orange or lilac to violet . . . 20
19 pores white, cream, yellow, yellowish green or greyish to brown . . . 22
20 pores red to orange: *Pycnoporus* (not incl.)
20 pores violet, lilac or apricot . . . 21
21 pores lilac to violet: *Hirschioporus* (not incl.)
21 pores pale orange to apricot: *Hapalopilus* (not incl.)
22 pores grey, yellowish, greenish to brown . . . 23
22 pores white, cream or yellow . . . 25
23 pores grey: *Bjerkandera* (part, incl. *B. adusta*)
23 pores yellowish to greenish, bruising/ageing darker, brown . . . 24
24 large thick bracket, splits easily; flesh with silky sheen; upper surface tomentose to shaggy; deciduous trees: *Inonotus*
24 bracket with broad basal stump; shaggy top surface with yellow margin; at base of conifers: *Phaeolus*
25 tube layer distinct from flesh . . . 26
25 tube layer not distinct from flesh . . . 30
26 pores bright yellow; upper surface yellow to orange; large, fleshy: *Laetiporus sulphureus*
26 pores white . . . 27
27 only on birch; small to large, shell- to kidney-shaped, white to pale brown; flesh white, spongy: *Piptoporus betulinus*
27 on birch with different characters *or* at base of other deciduous trees in large fan-like or fronded brackets, rarely with multiple rounded heads . . . 28
28 large to very large bracket at base of deciduous trees; fan- or frond-like to many-headed: *Grifola, Meripilus*
28 small brackets, not this combination of characters . . . 29
29 bracket white to cream: *Tyromyces* (not incl.)
29 bracket various shades of brown: *Polyporus* (part)
30 bracket thick, corky or woody; pores medium to large: *Trametes, Pseudotrametes*
30 bracket thin, leathery; pores small: *Coriolus*
31 pores dark greyish to brown; cap funnel-shaped, round with central stem; on burnt soil: *Coltricia perennis* (not incl.)
31 pores whitish and/or cap not round with central stem . . . 32

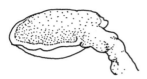

Ganoderma lucidum

Daldinia concentrica

truffle

32 shell- or kidney-shaped with lateral stem short or long; cap "varnished", chestnut to red-purple: *Ganoderma lucidum*
32 cap rounded, brownish, not "lacquered"; stem central to eccentric with base usually black, tomentose: *Polyporus* (part)

GLOBOSE OR STUD-LIKE FUNGI ON WOOD

1 flesh very hard, woody . . . 2
1 flesh like rubber . . . 3
2 medium to large hard black "ball"; when cut in section shows concentric bands or zones: *Daldinia concentrica*
2 small reddish brown balls scattered in large numbers; no concentric bands: *Hypoxylon fragiforme*
3 clusters of "studs" or thick "buttons" 2–5 cm across, 1–2 cm deep (globose, margin inrolled, when young), top may be depressed, cup-like; feels like soft rubber: *Bulgaria inquinans*
3 similar but with a short stem; more jelly-like: *Exidia recisa* (yellow-brown), *E. truncata* (black, not incl.) . . . 4

TRUFFLES

(only *Tuber*, true truffles, is included in this book)
1 skin smooth; interior with "chambers" . . . 2
1 skin smooth, spiny or warty; interior solid with coloured "veins" *or* convoluted veins or canals opening at surface . . . 6
2 pale whitish, or grey-brown to greenish . . . 3
2 ochre to deep brown; often with root-like rhizomorphs . . . 5
3 interior purplish brown; outside white to greenish; odour of cocoa; in beech woods: *Arcangeliella*
3 interior greyish to olive . . . 4
4 interior olive, hard; skin flaking, peeling: *Hysterangium*
4 interior pale grey with small open chambers: *Hymenogaster*
5 interior with open chambers; skin ochre to brownish: *Rhizopogon*
5 without distinct chambers; skin dark brown: *Melanogaster*
6 interior powdery; skin minutely warty or spiny, yellowish brown: *Elaphomyces*
6 interior not powdery; skin smooth or warty, not yellowish brown: *Tuber* and related genera

Guide
to
Species

Most mushrooms and toadstools are gill fungi or agarics (order Agaricales). These are fleshy, usually rather soft and usually with a distinct stem bearing the cap, on the underside of which the spores are produced on gills. On the basis of structure agarics are divided into families – Amanitaceae, Volvariaceae etc. The boletes and allied fungi, in which tubes usually take the place of gills, are separated in this book into the order Boletales (pp. 184–205).

AMANITACEAE

This widespread and very important family contains some of the best known poisonous and edible fungi. They are particularly abundant in warmer, southern temperate regions. *Amanita* species have white spores and a veil which forms various types of volva and usually a ring as well; the cap may be covered with fragments of veil. *Limacella*, the other and much smaller genus, has the same characteristic microscopic gill-structure but no volva, and always a viscid cap.

AMANITA

1 *Amanita pantherina* (Panther Cap): *ring not striate; 2–3 hoops or ridges at bulbous stem base; cap edge often striate*

Mixed woodlands. Autumn: uncommon to occasional. Cap 6–11 cm ($2\frac{1}{2}$–$4\frac{1}{2}$ in), convex then expanded; greyish brown to dull brown with whitish warts; margin striate when mature. Gills white, free, broad. Stem from rather squat to quite slender; white, smooth. Ring thin, not striate, white. Spores white, ovate, 9–12 × 7–9 µm, non-amyloid. Chemical test: sodium hydroxide on cap = orange-yellow. *Dangerous* but not usually fatal. In many districts rather rare; often confused with *A. excelsa*.

2 *Amanita phalloides* (Death Cap): *ring and bag-like volva; cap typically pale yellowish green to olive; specific chemical test*

Deciduous woods, especially oaks. Autumn. Occasional to frequent. Cap 5–12 cm (2–$4\frac{1}{2}$ in), convex then flattening, smooth; pale, almost white, in some forms; with faint radiating lines or fibrils. Gills broad, crowded, free, white. Stem white to pale greenish, often with faint horizontal banding, bulbous at base with large, white volva. Ring large, skirt-like, white, striate. Flesh white with sweet, sickly odour. Spores white, subglobose, 8–10 µm, amyloid. Chemical test: concentrated sulphuric acid on gills = pale lilac. *Deadly.*

3 *Amanita muscaria* (Fly Agaric): *unmistakable red cap and white or yellow warts; white stem, bulbous ridged base*

Under birch and pines. Autumn. Common. Cap 8–25 cm (3–10 in), bright scarlet or occasionally orange (in the U.S.A.

(4, 5, see over)

AMANITA (contd)

white and yellow forms are quite common), smooth; warts often washed off by rain. Gills white, free, broad, crowded. Stem tall, white. Ring large, skirt-like, white, often with yellow warts at margin. Spores white, elliptic, 9–11 × 6–8 μm, non-amyloid. *Dangerously poisonous*, and hallucinogenic.

4 *Amanita excelsa* (=spissa): *striate ring; bulbous, slightly scaly base; usually greyish or brown, not reddish or green*

In mixed woods, summer and autumn. Often abundant. Cap 6–15 cm (2½–6 in). Convex then expanded. Colour extremely variable with pearly white, grey and dark brown forms also common; smooth, with remains of veil sometimes as white warts but usually as white powdery patches. Gills white, crowded, free. Stem white to greyish, volva almost absent. Ring white, striate, skirt-like. Spores white, elliptic, 8–10 × 6–8 μm. Smell often absent, or potato-like. Edible, but should be avoided.

5 *Amanita rubescens* (Blusher): *strongly striate ring; no volva; similar to A. excelsa but always with reddish tints*

In woods generally. Summer to autumn. Very common. Cap 6–15 cm (2½–6 in), usually reddish brown (*very* variable) with warty or flaky patches. Stem bulbous, occasionally slight scaly zones at base. Flesh slowly reddens where damaged, especially in stem base where insect-infested. Spores ovate, 8–10 × 6–7 μm, amyloid. Edible when cooked but *poisonous* raw; best avoided.

1 *Amanita strobiliformis* (=solitaria): *large white cap with large polygonal warts; thick creamy veil remnants on cap margin and ring*

Open deciduous woodlands on calcareous soils. Autumn. Uncommon. Cap 8–20 cm (3–8 in). Gills white, crowded, free. Stem stout, often tall, and "rooting"; pointed base. Ring often torn or incomplete, texture of cream cheese. Spores elliptic, 10–12 × 8–10 μm, amyloid. Not poisonous but may be confused with other white dangerous species.

2 *Amanita caesarea* (Caesar's Mushroom): *unmistakable combination of orange cap, yellow stem and gills, and white volva*

Deciduous woods, especially oaks. Summer and autumn. Prefers warm temperature regions (the American type, possibly a distinct species, is more slender with slightly umbonate cap); not yet found in Britain. Cap 6–20 cm (2½–8 in), convex then expanded, clear orange; smooth, often with volval remnants adhering. Gills broad, crowded, free; egg- to chrome-yellow. Stem yellow, rather stout, bulbous with large sack-like white volva. Spores elliptic, 10–14 × 6–11 μm, non-amyloid. This delicious species, once highly valued by the Romans, is not easily confused with any other amanita.

(3, see over)

AMANITA (contd)

3 ***Amanita citrina*** *(=mappa): yellow or white cap; gutter-like volva; distinct smell of raw potato*

In mixed woods. Autumn. Common. Cap 5–9 cm (2–3½ in), pale lemon-yellow or pure white; usually with large flattened patches of veil. Gills white, free. Stem white, smooth, with "gutter" round bulb. Ring flaring, white or yellowish. Spores almost globose, 8–10 × 7–8 μm, amyloid. Very distasteful but not poisonous.

1 ***Amanita virosa*** (Destroying Angel): *white cap conical then expanding, large volva and shaggy stem; specific chemical test*

Mixed woods, usually at more northerly latitudes than most amanitas. Autumn. Occasional to frequent. Cap 6–12 cm (2½–4½ in), smooth, pure white and usually slightly sticky. Gills white, crowded, free. Stem tall, rather slender and often curved; usually rather rough and floccose, white, with large volva. Ring white, thin, often torn or missing. Flesh white with often a sickly, sweet odour. Spores white, globose 8–10 μm, amyloid. Chemical test: potassium hydroxide on cap = chrome yellow, distinguishing this species from other white amanitas. This *deadly poisonous* species contains many different toxins to which antidotes are only now being developed.

The related American species *A. bisporigera* is similar, but has a smooth stem, and only two spores per basidium, instead of four.

2 ***Amanita porphyria:*** *colours greyish with distinct violaceous tints; volvate base*

Under birches or conifers. Autumn. Occasional to frequent. Cap 4–10 cm (1½–4 in), convex to broadly umbonate, smooth, greyish brown with a violaceous tint; often with flat patches of white veil remnants adhering. Gills white, crowded, free. Stem white tinted lilac occasionally; bulbous with a white, bag-like volva. Ring pendent, flaring, striate and often torn; usually tinted strongly lilac or violaceous brown at margin. Spores subglobose, 7–10 × 6–8 μm, amyloid. Although not poisonous this species is extremely distasteful.

3 ***Amanita echinocephala:*** *sharp warts on cap and stem base: slightly greenish gills and spores*

In deciduous woods on calcareous soils. Autumn. Uncommon. Cap 7–20 cm (3–8 in), convex, white, discolouring brownish: smooth with numerous sharply pointed warts adhering but easily rubbed off. Gills white or very slightly greenish, free. Stem stout, bulbous with spiny warts around base. Ring pendent, white, striate. Spores white with a very pale greenish tint, 9–11 × 6–8 μm, amyloid. Its edibility is highly doubtful, and the species should be avoided.

AMANITA (contd)

1 *Amanita crocea:* cap and stem clear orange; stem shaggy-scaly, rather robust

Under birches. This more northerly species is frequent in Scotland and Northern Europe but rare elsewhere. Autumn. Very close to the more common *A. fulva* but stouter cap and stem a beautiful clear orange with no reddish brown; stem with horizontal bands of felty scales. Gills white. Volva pale orange. Spores white, subglobose, 9–13 × 8–11 μm, non-amyloid. Edible but not recommended.

2 *Amanita umbrinolutea:* cap olive-yellow to brown with light and dark zones

Deciduous woodlands, especially birches. Autumn. Uncommon. Very similar in shape to *A. fulva* and *A. vaginata* but quite distinct from either: larger and more robust, with distinctive colouring. Cap light olive-hazel with distinct lighter and darker zones, as illustrated; edge strongly striate-sulcate. Gills white. Stem pale cream-brown, slightly scaly. Volva large, bag-like, white or yellowish. Spores white, subglobose, 11–16 × 10–13 μm, non-amyloid. Edible but best avoided.

3, 4 *Amanita fulva* and *A.* vaginata (Grisette): cap orange-brown or grey; stem paler, with volva but no ring; often smooth patches left on cap

Mixed woodlands, especially birch. Summer and autumn. Common. Cap 4–10 cm (1½–4 in), ovate then soon expanded and slightly umbonate; smooth, dry and often with skin-like remnants of white veil-tissue; edge strongly striate-sulcate, orange-brown in *A. fulva*, pearly-grey in *A. vaginata*. Gills free, white. Stem tall and slender, tapering upward, white to pale brown or grey. Volva tall and sack-like, white outside, grey or brown within. Flesh fragile, crumbly, white. Spores white, globose, 9–11 μm, non-amyloid. Both are edible and good but best avoided, like all amanitas, to avoid confusion with poisonous species.

5 *Amanita inaurata* (=strangulata): large, with cap dark grey-brown or umber; stem tall, stout and shaggy with short warty volva and no ring; cap has angular warts, not smooth patches

Mixed woodlands. Autumn. Rather uncommon. Cap 6–15 cm (2½–6 in). Again very similar to *A. fulva* and *A. vaginata*, but larger and more robust with veil remnants angular, not smooth. Gills white. Stem tall, rather stout, grey-brown, rather tough with slightly shaggy horizontal bands of veil-tissue. Volva less bag-like than previous species, upper edge with thick angular warts readily breaking away. Spores white, globose, 10–13 μm, non-amyloid. Edibility uncertain; *possibly poisonous*, best avoided.

VOLVARIACEAE

This family has two principal genera, *Volvariella*, which is mainly tropical and has a volva, and *Pluteus*, without a volva, which is much more widely distributed with a very much larger number of species. They all share the common features of a pinkish spore deposit, gills free from the stem and with a unique arrangement of converging cells within the gills. A very large majority of the species in both genera occur on rotting wood or any rich organic matter such as straw in the tropics. Some species of *Volvariella* are cultivated for food and can even be purchased in cans sold under the title of Paddy-Straw Mushrooms.

VOLVARIELLA

1   ***Volvariella speciosa:*** *cap large and very sticky, usually grey-green; small volva; in rotting grass or straw, haystacks etc.*

Summer and autumn. Occasional to common. Cap 5–10 cm (2–4 in), very slightly fibrillose but otherwise smooth and very viscid in wet weather; pale grey-green to brown. Gills broad, pink. Stem white with short, thin white volva often almost missing or left behind when picked. Spores pink, elliptic, 13–18 × 8–10 μm. Edible but not highly recommended.

2   ***Volvariella murinella:*** *small to tiny, cap grey with jagged margin; grey volva*

In short grass, in fields and gardens. Summer and autumn. Frequent but often overlooked. Cap 2–4 cm (1–1½ in). Pale grey,

smooth. Gills broad, pink. Stem short, white with small grey volva. Spores pink, ovate, 6–8 × 4–5 μm. Edibility uncertain, but the species is too tiny to eat even if edible.

3 *Volvariella bombycina:* large, silky, pale cap and large volva; on dying or dead elms

Autumn. Uncommon. Cap 5–12 cm (2–4½ in), with a beautiful silky, fibrillose surface; pale yellow-green or almost white. Gills free, very broad and white, then soon deep pink. Stem white, bulbous with sac-like volva. Spores pink, ovate-elliptic, 7–9 × 5–6 μm. This lovely species is edible and delicious.

4 *Pluteus cervinus:* fibrillose cap (shades of brown) and stem; no volva; on deciduous wood

PLUTEUS

On stumps, logs and sawdust, all year. Very common. Cap 4–12 cm (1½–4½ in), convex, then soon expanded and slightly umbonate and often with a central depression, radially fibrillose, smooth; pale to dark brown. Stem white with darker fibrils, slightly bulbous. Gills broad, free, deep pink. Spores pink, ovate-elliptic, 6–8 × 4–5 μm. Edible but not highly favoured. On conifers the similar *P. atromarginatus*, with black gill edges, may be found.

PLUTEUS (contd)

1 ***Pluteus aurantiorugosus*** *(=coccineus): bright orange cap and stem; gills yellow, then pink.*

On decayed elms or ash. Autumn. Rare to occasional. Cap 2–5 cm (1–2 in), scarlet and globose when young, soon expanded and bright orange to yellow-orange; stem slender and long, not bulbous, pale orange-yellow. Spores pink, subglobose, 5 × 4–5 µm. One of the loveliest of all small toadstools, and not easily confused with anything else. Edibility uncertain; probably edible, but too small for use.

2 ***Pluteus salicinus:*** *blue-grey cap and lower half of stem*

Usually on willow trees and stumps, occasionally on beech. Autumn. Common. Cap 3–8 cm (1–3 in), smooth but finely fibrillose and minutely scaly at centre, grey with a blue-green tint. Gills broad, pink. Stem slender, pale blue-green fading to white above. Spores pink, elliptic, 8 × 5–6 µm. The colouring distinguishes this graceful species. Edible but not recommended.

3 ***Pluteus umbrosus:*** *dark brown, wrinkled and hairy-scaly cap, with fringed margin; stoutish stem also brown and woolly-scaly*

On decaying deciduous trees and stumps. Autumn. Occasional to frequent. Cap 3–8 cm (1–3 in), umber or sooty-brown; surface minutely wrinkled and roughened or reticulate with small, sparse woolly or hair-like scales; margin conspicuously fringed. Gills broad, crowded, pink, edge brown. Spores pink, elliptic, 6–7 × 4–5 µm. Edible but not recommended.

4 ***Pluteus lutescens:*** *dark brown cap with yellowish stem; gills pale yellow when young, then pink*

On rich soil. Autumn. Occasional. Cap 2–4 cm (1–1½ in), sooty-brown, smooth or slightly wrinkled, slightly umbonate when expanded. Stem slender, not usually bulbous; yellow especially in lower half. Spores pink, subglobose, 6–7 × 5–6 µm. A small but attractive species easily recognized when strongly coloured, less so in paler varieties. Edible but not worth eating.

LEPIOTACEAE

This family contains the beautiful Parasol mushrooms and many other attractive and fascinating smaller species. The gills are free from the stem, often with a distinct narrow "collar" between gills and stem, or adnate (*Cystoderma*). A well-developed veil is present, often forming a ring and leaving flakes on the cap and a floccose sheath below the ring. Spores are usually (all species included here) white and often dextrinoid. The family is rather southern in distribution.

MACROLEPIOTA

1 *Macrolepiota* (=Lepiota) *procera* (Parasol Mushroom): *open spaces; very large, with distinctive shape; darker bands on stem*

In fields and hedgerows, woodland margins. Summer and autumn. Common. Cap 10–25 cm (4–10 in), globular, soon flattening, with prominent umbo; pale biscuit-brown to hazel with thick flakes or scales as cap splits the universal veil. Gills crowded, broad, free, white to pale cream. Stem very tall and straight; pale brown with darker wavy zones. Ring large, thick, rather complex, high. Spores white, ovate, 13–17 × 8–11 µm, dextrinoid. One of our best edible species, almost unmistakable.

2 *Macrolepiota* (=Lepiota) *rhacodes* (Shaggy or Woodland Parasol): *usually in shade; no dark bands on stem; shaggy appearance*

In shade of woodlands, gardens etc. Autumn. Common. Cap 8–15 cm (3–6 in), soon flattening, not as prominently umbonate as *M. procera*; dull greyish brown with large, coarse brown flakes or scales. Gills broad, free, white to cream, staining red. Stem rather short, stout, smooth, *without* dark bands. Ring thick, double. Spores white, ovate, 8–11 × 5–7 µm, dextrinoid. Edible and delicious.

3 *Macrolepiota* (=Lepiota) *mastoidea*: *slender, delicate; cap strongly umbonate; pale colours and almost no scales*

In woodland clearings and margins. Autumn. Uncommon. Cap 5–10 cm (2–4 in), very pale biscuit-brown to almost white; finely roughened, granular, breaking up into small flakes. Gills white, free. Stem tall, slender, colour as cap, smooth. Ring high up, double. Spores white, 14–18 × 8–11 µm, dextrinoid. This beautiful species is edible and delicious.

LEPIOTA

4 *Lepiota friesii*: *warty cap and stem base; forked gills; repulsive odour and taste*

In deciduous woodlands, gardens. Autumn. Uncommon. Cap 6–10 cm (2½–4 in), conical then umbonate; pale yellow-brown, with pointed deep brown warts. Gills white, crowded, free, forked, often staining brown. Stem stout, rather short, pale brown with dark warty-scaly zones at base. Ring soft, skirt-like, white. Spores elliptic, 6–8 × 3–4 µm, dextrinoid. Inedible.

LEPIOTA (contd)

1　*Lepiota cristata* (Stinking Parasol): *cap centre red-brown with small concentric scales scattered towards margin, almost white; strong unpleasant smell of rubber or tar*

In short grass in fields, gardens, woodlands etc. Summer and autumn. Common. Cap 2–6 cm (1–2½ in). Gills white, crowded, free. Stem slender, pale brown. Ring very thin, often soon disappearing. Spores white, bullet-shaped, dextrinoid, 6–8 × 3–4 µm. *Possibly poisonous.* This small, rather attractively marked species is easily recognized by the unpleasant odour.

2　*Lepiota clypeolaria:* *fine zones of scales on cap with veil fragments at edge; slender stem woolly-floccose below*

In mixed woods. Autumn. Frequent. Cap 3–8 cm (1–3 in), convex with umbo, soon expanding; ochre to reddish brown, then cracking into rings of scales, finer at margin. Gills white, free, soft. Stem has ring-like zone. Odour slightly gas-like. Spores white, fusiform, 12–19 × 5–6 µm. Not recommended.

3　*Lepiota brunneoincarnata:* *small but stocky; cap minutely scaly with rose or wine-red flush*

In woodland clearings, gardens and roadsides. Uncommon to occasional. Autumn. Cap 3–5 cm (1–2 in), slightly umbonate, reddish brown with flush; densely squamulose at centre, less so at margin, scales darker brown. Gills cream or white, free. Stem pale reddish brown flushed below; squamulose-scaly up to the ring-like zone, bare above. Spores white, ovate, dextrinoid, 6–9 × 4–5 µm. Odour slightly fruity. *Poisonous.*

4　*Lepiota bucknalli:* *stem blue or lavender coloration; strong odour of coal-gas*

In deciduous woods on chalk or limestone. Autumn. Rare. Cap 2–4 cm (1–1½ in), slightly umbonate, white or pale grey, with a minutely granular surface; margin often with tiny toothed fringe. Gills white, free. Stem thin, lavender-blue or violet becoming dark blue below; surface mealy-granular. Ring absent. Spores white, bullet-shaped, 7–9 × 3–4 µm, dextrinoid. Edibility not known, *possibly poisonous.* This uniquely coloured species is easily recognized if you are lucky enough to see one!

5　*Lepiota leucothites* (=*Leucoagarius naucina*): *looks like an ordinary mushroom, but spores white and gills white then pale pink*

In fields, gardens, woodland clearings. Autumn. Uncommon. Cap 6–10 cm (2½–4 in), convex then expanding, white to pale cream; smooth and silky, slightly granular with age and breaking into small patches. Gills crowded, free. Stem white, base not usually bulbous. Ring thin, high on stem, white. Spores white, ovate, 7–9 × 5–6 µm. Edible but not recommended – can also be mistaken for a deadly amanita.

CYSTODERMA

1 *Cystoderma amianthinum:* granular cap and stem; yellow to orange-ochre colours; specific chemical test

In conifer or mixed heathy woodlands. Autumn. Common. Cap 2–5 cm (1–2 in), slightly umbonate; usually radially wrinkled, and minutely granulose. Gills white, adnate. Stem colour as cap; granulose up to ring zone. Spores white, 5–7 × 3–4 µm, amyloid. Chemical test: potassium hydroxide on cap cuticle = rust-brown. Edible but not recommended.

2 *Cystoderma carcharias:* main features as in the preceding species but pinkish-grey colour; specific chemical test

In conifer woodlands. Autumn. Frequent. Cap 2–4 cm (1–1½ in), pinkish-grey to flesh; margin denticulate. Gills white, adnate. Stem colour as cap. Ring usually more prominent than in previous species. Spores white, 4–6 × 3–4 µm, amyloid. Chemical test: potassium hydroxide on cap = no reaction, not brown. Edibility: not recommended.

AGARICACEAE

This distinctive family contains two important temperate genera. There are few *Melanophyllum* species but a great many *Agaricus*, including the best known and most valuable fungus in the world – the common cultivated mushroom. *A. brunnescens.* These "true" mushrooms have free gills, dark chocolate-brown spores, and usually a well-developed veil forming a ring.

MELANOPHYLLUM

3 *Melanophyllum echinatum* (=Lepiota haematosperma): gills blood- then brownish red; spores grey-brown, drying reddish

In woods and gardens on bare soil. Autumn. Uncommon. Cap 1–3 cm (½–1 in), minutely granulose; deep brown with margin denticulate. Stem pale reddish brown, finely granular; flesh red. Spores ovate, 4–6 × 2–4 µm. Edibility suspect.

AGARICUS

4 *Agaricus campestris* (Field Mushroom): in grassy fields; bright pink gills in young "buttons", then dark brown.

Grassy fields, roadsides. Summer and autumn. Common. Cap 5–10 cm (2–4 in), convex, slow to expand; smooth to slightly scaly-fibrillose; white to pale brown. Stem short, white with a thin ring often vanishing. Flesh white. Spores brown, 7–8 × 4–5 µm. One of the best edible species.

5 *Agaricus brunnescens* (=bisporus): on soil; fibrillose cap

Roadsides, gardens, manure heaps etc.; always on rich soil. Summer. Uncommon. Cap 5–10 cm (2–4 in), pale brown with fine radiating fibrils. Gills dull pink then dark brown. Stem rather longer than Field Mushroom, white with rather prominent ring. Spores brown, only two per basidium (hence *bisporus*), 6–8 × 5–6 µm. This excellent edible species is considered to be the ancestor of the Cultivated Mushroom.

AGARICUS (contd)

1 ***Agaricus variegans:*** *cap with flattened centre, broad brown scales; strong smell of rubber*

Mixed woodlands. Summer and autumn. Rare to occasional. Cap 5–15 cm (2–6 in), convex, flattened-depressed at centre; cream to pale ochre with broad, radiating brown scales. Gills pale pink, then dark brown. Stem slightly longer than cap diameter, white with woolly scales below. Ring large, pendent. Flesh, when cut, with strong smell like fresh rubber exactly as in *Lepiota cristata* (pp. 66–7). Spores 4–5 × 3–4 µm. A quite large species rather like a stout, brown-scaled Field Mushroom. Edible but not recommended.

2 ***Agaricus augustus:*** *large size, with cap yellow-orange, finely scaled; stem woolly*

Mixed woodlands. Summer and autumn. Frequent. Cap 6–20 cm (2½–8 in), convex, then flattening; bright tawny orange to brown with concentric rust-hazel fibrillose scales; edge with tattered remnants of veil. Gills broad, almost white, soon brown. Stem tall, stout, white; covered below ring with floccose, woolly scales. Ring large, pendent, skirt-like, scaly below. Spores ovate, 7–10 × 5–6 µm. Edible and delicious.

3 ***Agaricus arvensis*** (Horse Mushroom): *usually large and stout; cap yellowing; "cogwheel" ring on stem but no basal bulb and growing in open grass*

Very similar to *A. sylvicola* in form, colour, gills, ring etc., but larger and more robust and occurring outside woodlands, in fields, pathsides etc. Spores larger (7–10 × 4–5 µm). Edible and delicious.

4 ***Agaricus bitorquis*** *(==edulis, rodmannii): flat cap with thick flesh; narrow gills*

Common on roadsides and woodland edges; often forces up paving stones. Summer and autumn. Cap 6–15 cm (2½–6 in), soon noticeably flattened with margin slightly inrolled; smooth or slightly cracking; white, then discoloured pale brown. Gills very narrow and crowded, dull pink, then deep reddish brown. Stem short and stout; thick veil sheathing the base, rather like a volva. Often a thick narrow ring high on stem. Flesh thick, white, slowly becoming slightly brownish. Spores brown, subglobose, 5–7 × 4–5 µm. Edible and delicious.

5 ***Agaricus sylvicola:*** *small to medium size; yellowing cuticle; smooth bulbous stem; in shade of mixed woodlands*

Autumn. Common. Cap 6–12 cm (2½–4½ in), convex with broadly expanded margin; white then slowly yellowish with age or bruising. Gills grey-white (never bright pink), then soon dark brown. Stem rather tall and slender with bulbous base. Ring large, thin and pendent, of two distinct membranes, the lower often toothed like a cogwheel. Spores 5–6 × 3–4 µm. Smells of aniseed. Edible and delicious.

AGARICUS (contd)

1 ***Agaricus xanthodermus*** (Yellow Stainer): *vivid yellow flesh in stem base* and *below cap cuticle when cut*

Fields, gardens and hedgerows. Summer and autumn. Frequent to common. Cap 8–15 cm (3–6 in), white to greyish brown; cuticle often becoming coarsely cracked, even scaly; centre flattened to slightly depressed. Gills at first pale pink, then greybrown. Stem white, smooth, base bulbous. Ring large, pendent, double. Spores brown 5–6 × 4 μm. Smells of ink. *Poisonous to many*, edible to others, should not be experimented with. This is the only white mushroom that stains intense yellow in stem *and* cap.

2 ***Agaricus semotus***: *small, with purple colours in ivory cap with fine radiating lines*

Woodland clearings and margins. Summer and autumn. Uncommon. Cap 8–15 cm (3–6 in), pale hazel brown with darker, very fine scales overall; convex with flattened centre. Gills at first pale pink, then grey-brown. Stem white, smooth, bulbous; discolouring brown. Ring large, pendent, double. Flesh as in previous species turning chrome-yellow but at base of stem only. Spores brown, 4–6 × 3–4 μm. *Poisonous to some* people; to be avoided.

3 ***Agaricus placomyces***: *cap finely scaled, stem base (not cap) flesh goes bright yellow when cut; grows in woods*

Mixed woodlands. Autumn. Uncommon. Cap 3–7 cm (1–3 in), pale ivory with fine radiating purple fibrils; silky texture; often yellowing with age or bruising. Gills at first pale pink, then grey-brown. Stem tall, white, staining yellow below, base bulbous. Ring thin, flaring. Flesh white, only dull yellow when cut, never bright chrome-yellow. Spores 4–5 × 3 μm. Edibility not known; best avoided.

The closely related *A. phaeolepedotus* (overleaf), also in woodlands, rare, is similar in nearly all respects but has unchanging flesh.

4 ***Agaricus vaporarius***: *dull colours; scales thick, felty; stem very stout; often deeply rooted*

Woods or roadsides, in soil or leaf-litter, often in very large clumps with only the cap tops visible. Autumn. Uncommon. Cap 5–10 cm (2–4 in), dull grey-brown with thick, coarse scales; convex at first and only slowly flattening. Gills at first pale grey-pink, then dark brown. Stem very thick in proportion to length, often deeply rooting and pointed with thick shaggy veiltissue clinging to lower half. Ring thick, felty, often torn or missing. Flesh slowly reddening when cut. Spores 6–7 × 4–6 μm. Edible but not highly recommended.

AGARICUS (contd)

1 ***Agaricus sylvaticus*** (Scaly Wood Mushroom): *cap with brown scales; stem white; flesh reddening*

In mixed woodlands in leaf-litter or needles of pines. Autumn. Common. Cap 5–9 cm (2–3½ in), convex then flattened, pale brown or almost white in some forms (especially under pines) with darker brown to reddish brown scales. Gills crowded, rather narrow, pale pinkish then soon brown. Stem rather slender, swollen at base, white; smooth to slightly scaly below the ring. Flesh when cut white then soon pale to bright pinkish red especially in cap and outer flesh of stem. Spores brown, ovate, 5–6 × 3–4 µm. Edible and excellent.

The similar *A. langei* has a tall, stout, *non*-bulbous stem, and cap strongly reddish brown; its ring soon vanishes.

2 ***Agaricus macrosporus:*** *cap often very large, white then dull ochre, discoloured; stem short and stout*

In open pasture land. Summer and autumn. Occasional. Cap 10–25 cm (4–10 in), convex then slightly expanded; smooth at first then slightly scaly at margin. Gills whitish-grey then slowly brown; crowded, rather narrow. Stem stout, rather short, white; rough, woolly-scaly below the large, soft ring. Flesh white to slightly orange-pink when cut. Spores elliptic, brown, 8–12 × 5–7 µm. This good edible mushroom becomes rather ill-smelling with age.

3 ***Agaricus squamuliferus:*** *stocky, white, finely scaly mushroom with flesh turning bright red*

At woodland margins and clearings. Autumn. Uncommon. Cap 4–10 cm (1½–4 in), convex then expanding; white becoming brownish with age, surface disrupting into fine scales. Gills crowded, dull pink then brown. Stem rather stout, white, slightly woolly-scaly. Ring pendent, thin. This fungus is almost unique in being white (not brown, scaly) with flesh of cap and stem rapidly turning bright red when cut. Spores ovate, brown, 6–8 × 4–5 µm. Edible and delicious.

4 ***Agaricus phaeolepidotus:*** *cap smooth, silky with faint, adpressed scales; stem very smooth; flesh colour unchanging*

In deciduous woodlands on grassy soil, on straw or compost. Autumn. Uncommon. Cap 6–10 cm (2½–4 in), convex with a flattened central "boss"; pale cinnamon-brown with scales darker. Gills crowded, pale greyish pink then brown, never bright pink. Stem long, smooth, never scaly, white. Flesh white, unchanging, or slightly very pale ochre-yellow at base. Spores 5–6 × 3–4 µm. As a member of the *A. xanthodermus* group, which is mildly poisonous, this species is best not eaten. It somewhat resembles some red-staining, scaly wood mushrooms, e.g. *A. sylvaticus* (see above), but differs in its smoother, silky cap and stem and unchanging flesh.

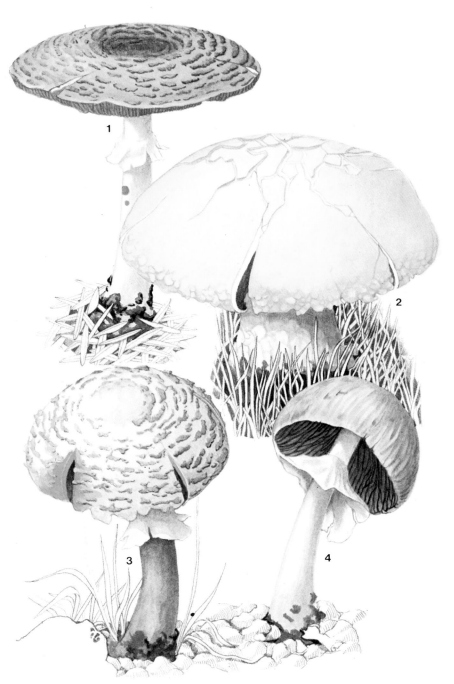

COPRINACEAE

This distinctive family has very dark spores, and gills from free to adnexed or adnate and often (*Coprinus*, the Ink Caps) deliquescing when mature. There is usually a well-developed veil and occasionally a ring. Many grow on dung (when they may vary considerably in size) and rotten wood. They are on the whole rather delicate, small and dull-coloured.

COPRINUS

1 *Coprinus atramentarius:* cap grey and slightly scaly; stem with ridged bulb at base, sometimes rooting

In clusters in gardens, roadsides, near old stumps or on bare soil. Summer and autumn. Common. Cap 2–6 cm (1–2½ in), broadly conical-ovate then expanding to bell-shaped; pale brownish grey, darker and slightly scaly at centre. Gills narrow, crowded, free; pale grey, soon black and dissolving. Stem white, elongating as cap expands, tapering upwards. Spores black, ovate, 7–10 × 5–6 µm. This is quite tasty and apparently *harmless, unless consumed with alcohol*, when rather alarming symptoms occur of facial flushing, nausea and palpitations.

2 *Coprinus comatus* (Shaggy Mane, Shaggy Ink Cap, Lawyer's Wig); *unmistakable tall, white, shaggy cap; elegant slender stem with ring; black, dissolving gills*

On grassland, roadsides, playing fields etc., especially on turf over soil recently disturbed. Common. Cap 5–15 cm (2–6 in) high, cylindrical then narrowly bell-shaped; white to pale brown at centre; very shaggy with woolly scales. Gills crowded, narrow, almost free; white, then pink, then black, dissolving from margin until almost entire cap is gone. Stem tall (10–20 cm), smooth, white. Ring narrow, easily movable. Spores blackish purple, elliptic, 13–14 × 7–9 µm. Delicious and delicately flavoured when young.

3 *Coprinus micaceus* (Glistening Ink Cap): *cap date-brown, grooved; with scattered "crystals" on surface when young*

In large clumps on deciduous stumps. Summer and autumn. Common. Cap 3–6 cm (1–2½ in), yellow to date-brown, ovate-conical, grooved. Gills white, then almost black, slightly autodigesting. Stem medium to long, thin, white and smooth. Spores lemon-shaped, 7–12 × 6–7 µm. Edible but small.

4 *Coprinus lagopus:* very tall stem; cap woolly-hairy, white; usually solitary

Woodlands, on woody debris. Autumn. Frequent. Cap 2–5 cm (1–2 in), cylindrical-ovate, white, woolly-hairy, then rolling upward, papery, as the grey gills shrink and contract. Stem short at first. Spores 10–12 × 5–8 µm. Edible but not worth considering.

(5, see over)

COPRINUS (contd)

5 *Coprinus disseminatus* (Fairies' Bonnets): *caps very small, strongly grooved in large clusters (often of hundreds)*

On stumps of deciduous trees. Autumn. Common. Cap 1–2 cm ($\frac{1}{2}$–1 in), ovate, then expanded; thin, membranous, grooved (more strongly than *C. micaceus*); grey or pale ochre, not autodigesting. Gills white, then grey, Stem thin, short, finely hairy. Spores 8–11 × 4–5 µm. Not poisonous, but too small.

1 *Coprinus niveus*: *beautiful chalk-white toadstools on dung*

Always on cow or horse dung. Summer and autumn. Common. Cap 1–4 cm (8–18 in), ovate, soon expanding to bell-shaped, texture chalky-scurfy. Gills grey, then black, soon autodigesting. Stem long, thin, white, texture as cap; no ring. Spores lemon-shaped, 12–18 × 10–12 µm. Edibility unknown.

LACRYMARIA

2 *Lacrymaria velutina* (=*Hypholoma velutinum*) (Weeping Widow): *yellow-brown fibrillose cap; black weeping gills*

On road- and pathsides, in gardens and fields. Summer and autumn. Very common. Cap 4–10 cm ($1\frac{1}{2}$–4 in), convex, often with a low umbo, covered (especially when young) with woolly fibrils, veil remnants especially at margin; pale clay to ochre-brown, sometimes pale orange. Gills crowded, sinuate, dark brown to black; edges white with small droplets when fresh and rapidly developing. Stem quite thick, with woolly fibrils up to

ring-like zone. Spores almost black, lemon-shaped and warty, 10–12 × 6–7 µm. Edible.

3 _Psathyrella caput-medusae:_ _cap pale with broad, dark scales; lower stem scaly; on or around conifer stumps_

Autumn. Rare to occasional. Cap 3–6 cm (1–2½ in), convex then expanded, almost white with deep brown centre and scales on outer half. Gills adnate, grey then dark brown. Stem stout, white, smooth above the thick, flaring ring. Spores 10–12 × 4–5 µm. Edibility uncertain.

4 _Psathyrella candolleana:_ _almost white caps, very fragile; in tufts on or by deciduous stumps_

Summer and autumn. Common. Cap 3–8 cm (1–3 in), convex, then expanded, pale buff to white; smooth, but tiny tooth-like veil remnants at margin. Gills adnate, pale greyish lilac, then deep brown. Stem thin, very fragile, white. Spores 7–8 × 4–5 µm. Too fragile, even if its edibility is proven.

5 _Psathyrella hydrophila:_ _cap changing colour on drying; faint ring zone on stem; in clusters on deciduous stumps_

Autumn. Common, Cap 3–8 cm (1–3 in), convex, soon almost flat; smooth, date-brown; rapidly drying; pale buff, edge bears cobwebby veil. Gills adnate, pale, then deep brown. Stem white, smooth. Spores 5–7 × 3–4 µm. Edibility uncertain.

PSATHYRELLA (contd)

1 *Psathyrella multipedata:* tall thin stems all coming from one common "root"

Pathsides in woods. Autumn. Uncommon. Cap 1–3 cm ($\frac{1}{2}$–1 in), conical, then expanded; date-brown when moist, drying out clay-buff; margin striate. Gills adnate, grey, then dark purple-brown. Stem long, thin, fused at base with many others. Spores 6–10 × 4–5 μm. Edibility uncertain.

2 *Psathyrella conopilea:* usually solitary; tall, thin, with conical brown cap and deep brown gills; red-brown hairs on cap visible under lens distinguish this from similar species

On soil in woods and gardens. Autumn. Common. Cap 2–5 cm (1–2 in), date-brown when moist, pale buff when dry. Gills adnate, grey-brown, then dark purple-brown. Stem long, thin, white and smooth, no ring. Spores 12–16 × 7–8 μm. Edibility uncertain.

PANAEOLINA

3 *Panaeolina (=Panaeolus) foenisecii:* common in garden lawns; small, often zoned cap

In lawns, fields and roadsides. Summer and autumn. Very common. Cap 1–3 cm ($\frac{1}{2}$–1 in), globose to convex, dark brown when moist, pale cinnamon on drying – centre often dries first, producing a bicoloured or zoned appearance. Gills adnate, pale to dark brown, colouring unevenly ("mottled"). Stem thin, smooth, brown. Spores black, lemon-shaped, warty, 12–15 × 7–8 μm. Not recommended; *has caused poisoning.*

PANAEOLUS

4 *Panaeolus semiovatus (=Anellaria separata):* on dung; cap slimy, grey-brown, bell-shaped; distinct ring

On horse dung in fields. Summer and autumn. Common. Cap 3–6 (–8) cm (1–3 in), globose, then bell-shaped; smooth, pale greyish tan or flesh colour; viscid (tacky) when moist. Gills adnate, mottled greyish, then black. Stem tall to very tall (6–16 cm), rigid, smooth, pale tan below and white above erect membranous ring (ring sometimes torn or absent). Spores black, 18 × 10 μm. Edibility uncertain; best avoided.

5 *Panaeolus sphinctrinus:* conical grey cap with margin "teeth"

In fields on or around dung. All year. Common. Cap 2–4 cm (1–1$\frac{1}{2}$ in), conical, smooth; grey to greyish brown, margin with distinct tooth-like (denticulate) fringe of veil remnants, particularly when young. Gills adnate, "mottled", greyish, then almost black, edge white. Stem long, thin, smooth, colour of cap; white-powdered at apex. Spores black, lemon-shaped, 14–18 × 10–12 μm. Edibility uncertain: some related species are poisonous, sometimes hallucinatory. A very distinctive species if the colour and the small "teeth" are both typical; occasional specimens with the teeth missing and/or of darker colour may prove misleading.

STROPHARIACEAE

This family contains some brightly coloured and beautiful fungi, including some blue and green species – both rare colours in toadstools. Nearly all have brown spores and are found on dead wood, wood chips or soil containing woody organic matter. Very few are considered edible, with only a couple being cultivable, while some (particularly in *Psilocybe* and *Stropharia*) are decidedly poisonous or contain hallucinogenic toxins. Other distinguishing characteristics are the adnate to sinuate gill attachments, and, usually, a veil present as a marginal veil and/or ring. Genera included here are *Hypholoma*, *Stropharia* and *Psilocybe*.

HYPHOLOMA

1 *Hypholoma fasciculare* (Sulphur Tuft): *cap and (when young) gills sulphur-yellow; on deciduous wood; bitter taste*

In spectacular large clumps, often in hundreds, on stumps, logs or diseased deciduous trees. Summer and autumn. Very common. Cap 4–8 cm (1½–3 in), convex to slightly umbonate; smooth, bright sulphur-yellow with a tint of orange at the centre; margin with traces of veil remnants. Gills sinuate, sometimes adnate; yellow then soon greenish and finally dark purple-brown. Stem long, fibrous, pale yellow, brownish at base with a ring zone above. Spores purple-brown, 5–7 × 3–5 μm. Flesh very bitter and inedible.

2 *Hypholoma sublateritium* (Brick Caps, Red Caps): *cap and stem large, robust, with reddish (never sulphur) colours; on deciduous trees, usually later than other species*

In large clumps on deciduous trees and stumps. Late autumn. Frequent. Cap 4–10 cm (1½–4 in), convex, then flattened; smooth, rich brick-red with paler margin; often with yellowish veil remnants at edge. Gills sinuate, pale yellow, then greyish lilac. Stem long but stouter than previous species, fibrous, pale brick-red, and with ring zone. Spores rich brownish lilac, 6–7 × 3–5 μm. Usually considered inedible, although definitely eaten in the U.S.A. – there are possibly two differing strains. This large, attractively coloured species usually appears later than the other hypholomas mentioned here.

3 *Hypholoma capnoides* (Conifer Sulphur Tuft): *cap pale ochre-yellow; gills whitish when young; mild taste; on conifers*

Almost identical to the previous species, but differing in its habitat on conifers, the duller, more ochre yellow of the cap, the gills whitish at first then greyish lilac, not yellow-green, and the mild taste. Spores 7–9 × 4–5 μm. Edible, but not recommended.

HYPHOLOMA (contd)

1 Hypholoma udum: *stem tall and thin; cap umbonate, reddish ochre; spores purple-brown*

In bogs, marshes, etc., in sphagnum moss. Autumn. Common. Cap 1–2 cm ($\frac{1}{2}$–1 in), convex to bell-shaped with an umbo at the centre; dull reddish ochre, smooth, slightly viscid. Gills adnexed, pale yellow-olive, then dark brown. Stem long, thin, colour as cap. Spores 14–18 × 6–7 µm. Typical of a number of small slender marsh-loving species of *Hypholoma* difficult to tell apart. Edibility doubtful; best avoided.

PSILOCYBE

2 Psilocybe squamosa (Hypholoma squamosum): *cap convex with tiny scales; tall stem scaly below ring; spores purple-brown*

On soil containing wood chips on fallen twigs. Autumn. Uncommon to occasional. Cap 2–6 cm (1–2$\frac{1}{2}$ in), convex, smooth with tiny scattered brown scales; pale ochre-yellow. Gills adnate, pale yellow, soon purple-brown. Stem tall, thin, pale ochre with scaly bands below, smooth above. Ring flaring, often torn. Spores 11–15 × 6–8 µm. Edibility suspect.

3 Psilocybe semilanceata (Liberty Cap): *very distinctive, sharply pointed cap; pale yellow colours*

In grass in fields, gardens, roadsides etc. Summer and autumn. Common. Cap 1–2 cm ($\frac{1}{2}$–1 in) high, margin inrolled when young; with acute conical point; smooth, viscid, cuticle easily separable; pale ochre-yellow to buff with greenish tints at margin. Gills adnate, purple-brown. Stem tall, thin, wavy, white above becoming pale yellow below; some specimens turn blue at base when picked. Spores purple-brown, 12–14 × 7 µm. Not suitable for culinary purposes as it contains varying amounts of hallucinogenic chemicals.

STROPHARIA

4 Stropharia aeruginosa (Verdigris Agaric): *unmistakable vivid blue-green slimy cap; robust stature*

In grass in woods, gardens and hedgerows. Summer and autumn. Frequent. Cap 3–8 cm (1–3 in), convex with a low umbo, smooth and very viscid; rich blue-green soon fading to pale yellow-green; white woolly veil remnants on cap when young. Gills broad, adnate, purple-brown. Stem stout, white or pale green, woolly-scaly below the thin narrow ring, smooth above. Spores purple-brown, 7–9 × 4–5 µm. Edibility suspect.

5 Stropharia pseudocyanea: *colours paler and bluer than Verdigris Agaric; stature delicate and slender; cap smaller*

In grass in woods and fields. Autumn. Uncommon. Cap 1–3 cm ($\frac{1}{2}$–1 in), convex, flattening with very slight umbo, smooth, viscid; pale blue with less green than *S. aeruginosa*, fading rapidly to yellowish. Gills adnate, purple-brown. Stem long, slender, pale yellow with blue in lower half. Ring thin, often almost absent. Spores purple-brown, 8–9 × 4–5 µm. *Possibly poisonous.*

STROPHARIA (contd)

1 *Stropharia semiglobata* (Dung Roundhead): *distinctive hemispherical cap and tall thin stem; on dung*

On dung or dung-enriched grass. Summer and autumn. Common. Cap 1–4 cm ($\frac{1}{2}$–$1\frac{1}{2}$ in), hemispherical to slightly umbonate, not flattening with age; pale yellow to ochre, smooth, viscid. Gills broadly adnate, almost triangular, dark purple-brown. Stem long, thin, pale yellowish; below ring viscid, then dry and shiny. Ring very thin, narrow, sometimes imperfect or absent. Spores dark-brown, 18–20 × 10 µm. Not edible. This common and easily recognized species is very variable in size.

2 *Stropharia ferrii* (=rugosoannulata): *brick-red cap and smooth stem; preference for wood-chips*

On soil in flower beds and roadsides. Autumn. Rare to occasional. Cap 5–20 cm (2–8 in), convex, then soon flattened; pale to rich brick-red, sometimes purplish; smooth, dry with fine innate fibrils, edge often splitting. Gills dark purple-brown, sinuate-adnate. Stem tall, stoutish, white and quite smooth when mature but slightly woolly when young. Ring thick, edge splitting. Spores 10–14 × 6–8 µm. This large, impressive species looks much like a mushroom (*Agaricus* species) at first sight. It is edible and is cultivated in parts of Europe.· Common in America, less so in Europe and very infrequent to rare in Britain.

3 *Stropharia coronilla*: *mushroom-like shape with pale yellow cap; grooved ring*

In fields, pastures, and among dune-grasses. Summer and autumn. Frequent. Cap 2–4 cm (1–$1\frac{1}{2}$ in), convex, then expanded, pale straw to ochre-yellow; smoóth, not viscid as in previous species or only slightly so in wet weather. Gills adnate, almost sinuate, purple-brown. Stem short, stoutish, white or pale yellow, especially below ring. Ring thicker and more prominent than in *Stropharia semiglobata*, radially grooved above. Spores purple-brown, 8–9 × 4–5 µm. Edible but not recommended: confusion possible with other species.

4 *Stropharia hornemannii* (=depilata): *large glutinous yellow-brown cap; flesh smells unpleasant*

On fallen twigs or sawdust of conifers. Autumn. Uncommon to locally frequent. Cap 5–15 cm (2–6 in), convex, then flattened with a low umbo, very viscid, glutinous; pale ochre-yellow to darker brownish yellow, sometimes flushed violet when young. Gills adnate to sinuate, purple-brown. Stem medium to long, stoutish, white or pale straw-yellow; woolly-scaly up to the ring. Ring large, pendent. Spores purple-brown, 11–14 × 6–8 µm. Not edible, *probably poisonous*.

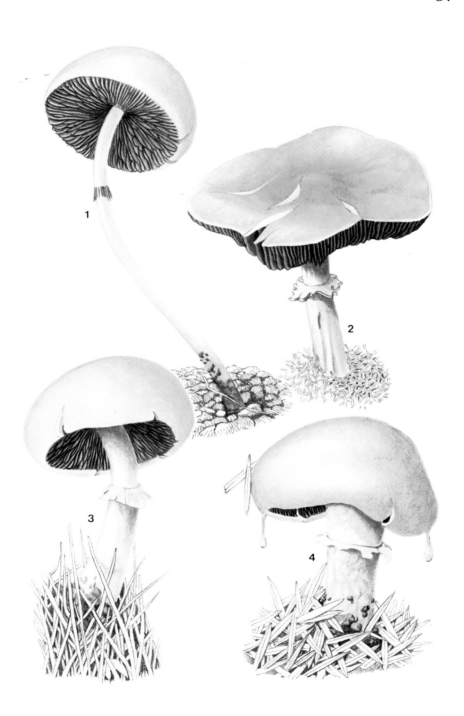

CORTINARIACEAE

One of the largest and most distinctive families of agarics, this group contains species of all sizes, habits and appearances, from robust, clustered tree-dwellers to tiny, delicate swamp fungi. *Cortinarius* is perhaps the largest genus in the world, with several hundred species. Many are exceptionally beautiful and many unfortunately are rare. All the genera included here have a curtain-like veil (cortina), tawny orange or brown (often rust-brown) spores, and other anatomical features more difficult to observe.

PHOLIOTA

1 ***Pholiota aurivella:*** *in the upper parts of trees (usually beech); large glutinous cap, dry stem*

Clustered or solitary. Autumn. Common. Cap 5–18 cm (2–7 in), convex, then soon flattened; rich golden yellow or orange, with darker adpressed scales, also glutinous. Gills broad, crowded, sinuate; pale yellow then rust-brown. Stem with ring zone, stout, usually horizontal, often long; pale yellow-orange, not slimy, extremely tough and fibrous. Spores elliptic, rust-brown, 8–9 × 4–6 μm. Flesh tough. Not edible. A large, easily recognized species. The similar *P. adiposa*, at base of trees, has cap *and* stem slimy.

2 ***Pholiota gummosa:*** *very glutinous pale greenish yellow cap; stem white, floccose and sticky*

On stumps but more often on buried wood, in clusters. Summer and autumn. Frequent. Somewhat similar in shape to the more common Charcoal Pholiota, but with 3–5 cm (1–2 in) cap distinctly glutinous, pale greenish yellow, with white woolly scales soon vanishing. Stem white, then yellow in lower half, rust-brown at base, also floccose. Spores rust-brown, elliptic, 5–7 × 3–4 μm. Edible.

3 ***Pholiota alnicola:*** *in clusters at base of birch or willow; in late autumn; cap bright yellow with almost dry cuticle; bitter taste*

Frequent. Cap 4–12 cm (1½–4½ in), globose, then convex with low umbo; smooth, only slightly viscid; bright primrose to sulphur-yellow, sometimes greenish at edge. Gills adnate, pale yellow, soon ochre to pale rust-brown. Stem long, often curved or twisted, fibrous, pale yellow above becoming rust-brown below. No ring, but faint white veil at stem apex and on cap margin. Flesh bitter, tough, strong smell. Spores elliptic, rust-brown, 7–9 × 4–6 μm. Not edible.

4 ***Pholiota highlandensis*** *(=carbonaria)* (Charcoal Pholiota): *always clustered in burnt woodland; cap small, orange-brown*

Often in large numbers. Summer and autumn. Common. Cap 2–6 cm (1–2½ in), convex, viscid when moist, dry and shining in dry weather; rich orange-yellow to ochre. Gills adnate, pale

PHOLIOTA (contd)

clay then dull brown. Stem dull yellow or orange, browner at base; viscid below ring zone, slightly scaly. Spores 7–8 × 3–4 µm. Not considered edible.

1 *Pholiota squarrosa* (Shaggy Pholiota): *large scaly yellow cap and stem; gills rust-brown*

At base of broadleaved trees, often in clumps. Autumn. Common. Cap 4–12 cm (1½–4½ in), convex then soon expanded with a low umbo; yellow-ochre to tawny orange, covered with hazel-brown recurved scales. Gills arcuate-decurrent, straw-yellow then soon rust-brown. Stem variable in length, tough, fibrous; smooth above thin, torn ring, scaly below; colour as cap. Spores rust-brown, 6–8 × 4 µm. Apparently edible but very tough and indigestible.

2 *Pholiota flammans:* *cap bright yellow with paler scales; on conifer stumps and logs*

Usually in small clumps, or solitary. Autumn. Occasional to frequent. Cap 3–8 cm (1–3 in), convex then expanded with slight umbo, brilliant chrome to tawny yellow with paler recurved scales. Gills crowded, adnexed, pale sulphur-yellow then soon rust-brown. Stem bright yellow with recurved scales below the thin ring, smooth above. Spores rust-brown, 4–5 × 2–3 µm. Edibility doubtful; best avoided.

3 *Pholiota lubrica:* *brick-red viscid cap, paler stem; often on ground*

On rich soil, buried wood, near stumps. Autumn. Uncommon. Cap 3–7 (1–3 in), convex with obtuse umbo, then flattening; smooth, viscid, sometimes with pale, filmy scales; brick-red tawny to cinnamon, paler at margins. Gills adnate, pale whitish then pale clay. Stem with ring zone, fibrillose, whitish then darkening to reddish. Spores rust-brown, elliptic, 6–7 × 3–4 µm. Edible but not recommended.

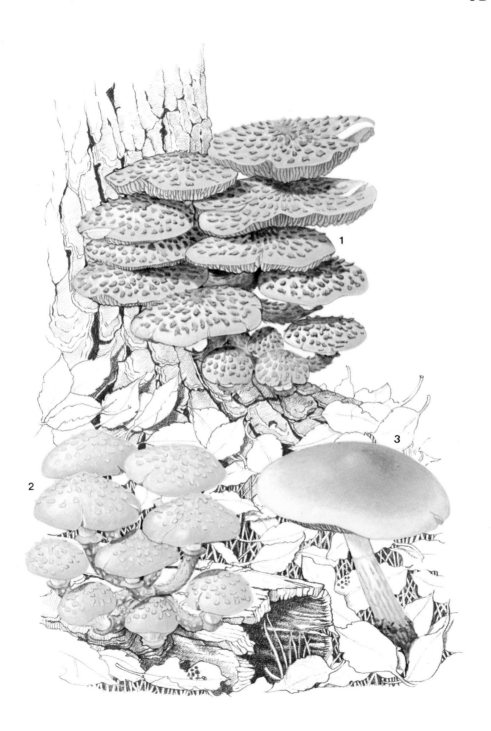

ROZITES

1 Rozites caperata: *pale clay-coloured cap with white "frosting"; white stem; clay-coloured gills*

In mixed woodlands with heathy vegetation on acid soils. Autumn. Occasional. Cap 5–10 cm (2–4 in), convex then soon expanded with umbo; pale ochre-yellow or brown cap with a coating of white, finely scattered veil-like minute crystals. Gills pale clay-brown, adnexed. Stem white, smooth or with thin belts of white veil; volva-like base. Ring thin, double, often torn, sometimes missing. Spores brown, finely warted, 11–14 × 7–9 μm. Edible but not highly recommended.

INOCYBE

Inocybe is a very distinctive genus usually easily recognized by the radially fibrillose cap and fibrous stem, dull cigar-brown spores and often pungent smell. Most are *dangerously poisonous.*

2 Inocybe perlata: *dull ochre cap with inturned margin; stout stem*

Edges of woodlands, roadsides etc. Autumn. Uncommon. Cap 3–10 cm (1½–4 in), soon expanded with low umbo, smooth, with an abruptly incurved margin; pale ochre streaked with slightly darker fibrils. Gills crowded, adnate, cigar-brown with white edge. Stem rather stout (1–1½ cm); white but darker below; finely fibrillose, not bulbous. Spores dull brown, elliptical, 9–12 × 6–8 μm. As with all inocybes, *possibly dangerously poisonous.*

3 Inocybe patouillardii: *white bell-shaped cap with split margin and white stem, all parts bruise or age bright pinkish red; olive-brown gills*

In deciduous woods, especially beech on chalky soil. Summer and autumn. Occasional. Cap 3–8 cm (13 in), conical then campanulate, smooth or finely fibrillose; margin of cap often split; white. Gills whitish then pale olive-brown. Stem stout, tough, fibrous, white, usually not bulbous. Spores dull-brown, bean-shaped, 10–13 × 5–7 μm. Odour strong, fruity or pungent. *Dangerously poisonous, even fatal.*

4 Inocybe geophylla (Common White Inocybe): *small white, umbonate cap and white stem; clay-brown gills; sickly smell*

In mixed woodlands. Autumn. Common. Cap 1–3 cm (½–1 in), convex then soon expanded and umbonate, smooth, silky; white, slightly discoloured when older. Gills cream at first. Stem thin, smooth, white, not bulbous. Spores dull brown, 8–10 × 5–6 μm. Odour sickly, sweetish. *Poisonous.*

5 Inocybe geophylla var. **lilacina:** *identical in all respects to the preceding except for the beautiful clear lilac coloration overall*

INOCYBE (contd)

1 *Inocybe pudica:* broadly umbonate cap, white flushed pink; under conifers; different spore size from the similar I. patouillardii

Cap 3–5 cm (1–2 in), convex then expanded and umbonate, silky fibrillose surface; white flushing pink to brown. Gills pale clay then dull brown with reddish tints. Stem rather stout, slightly bulbous, white staining reddish. Spores dull brown, bean-shaped, 8–9 × 4–6 μm. Odour rancid or mealy. *Poisonous.*

2 *Inocybe hystrix:* brown, sharply scaly cap and stem

Beech woods. Autumn. Uncommon. Cap 2–5 cm (1–2 in), convex with slight umbo; dark brown, covered with pointed erect scales. Gills crowded, whitish then pale clay, edge white. Stem brown as cap with recurved scales, apex paler, without scales; base not bulbous. Spores dull brown, pip-shaped, 9–10 × 4–5 μm. *Poisonous.*

3 *Inocybe jurana:* cap, stem and flesh with distinct flush of deep wine-purple

Mixed woodlands, often deeply sunk in soil. Autumn. Uncommon. Cap 4–8 cm (1½–3 in), conical then expanded with blunt umbo; fibrillose, pale brown with dark wine-purple flush. Gills crowded, whitish then olive-brown, edge white. Stem stout, fibrous, usually not bulbous; white flushed with darker wine-purple fibrils especially at base. Flesh also with flush. Spores dull brown, bean-shaped, 10–15 × 5–7 μm. Edibility uncertain, *probably poisonous.*

4 *Inocybe griseolilacina;* brown tomentose cap with lilac stem

In deciduous woods. Autumn. Occasional. Cap 1–3 cm (½–1 in), convex with papillate (nipple-shaped) umbo; pale brown, finely shaggy tomentose. Gills white then dull brown. Stem thin, pale lilac, finely floccose, not bulbous. Spores dull brown, elliptic, 8–10 × 5–6 μm. *Poisonous.*

5 *Inocybe maculata:* conical, fibrillose brown cap with patches of white veil; stem white then soon discoloured brown

In deciduous woods and on pathsides. Autumn. Frequent. Cap 4–8 cm (1½–3 in), usually sharply conical; fibrous, cigar-brown with pale filmy veil patches at centre. Gills pale whitish then cigar-brown. Stem fibrillose, non bulbous. Spores dull brown, bean-shaped, 9–11 × 4–6 μm. Odour rather pungent, strong. *Poisonous.*

INOCYBE (contd)

1 Inocybe cookei: *straw-yellow umbonate cap; marginate white bulb on stem*

On soil in mixed woods. Autumn. Frequent. Cap 2–5 cm (1–2 in), expanded with distinct umbo; smooth at centre, fibrillose and cracking at margin; pale straw-yellow. Gills whitish then dull clay-brown. Stem above bulb paler than cap, smooth. Spores dull brown, bean-shaped, 7–8 × 4–5 µm. *Poisonous.* The rather similar *I. praetervisa* has irregular, nodulose spores.

2 Inocybe asterospora: *obtusely umbonate cap; marginate bulb to stem; distinctive spore shape*

In mixed woods, by paths on soil etc. Autumn. Frequent. Similar in most respects to the more common *I. napipes* (below) but cap less sharply umbonate, fibrillose with paler flesh exposed between the brown fibres; stem with a distinctly marginate bulb; and spores almost star-shaped (not oblong) with 5–8 conical knobs, 9–12 × 8–10 µm. *Poisonous.*

3 Inocybe bongardii: *pale brown cap; tall stem, white flushed reddish; strong fruity odour, especially of ripe pears*

At edges and pathsides of mixed woods. Summer and autumn. Uncommon. Cap 3–7 cm (1–3 in) convex to bell-shaped, fibrillose with scattered minute scales; pale brown with a flesh tint. Gills broad, whitish, soon olive-cinnamon. Stem tall, rather stout, fibrillose. Flesh reddish. Spores dull brown, bean-shaped, 10–13 × 5–7 µm. *Poisonous.*

4 Inocybe napipes: *acutely umbonate, brown fibrillose cap; non-marginate bulb to stem*

In damp leaf-litter and soil in deciduous woods, especially birch. Autumn. Common. Cap 2–5 cm (1–2 in), conical at first; fibrillose, dark chestnut-brown. Gills whitish then dull brown. Stem tapering above, swollen below with a non-marginate bulb; fibrillose, brown but paler below. Spores oblong, nodulose, with 5–6 blunt knobs, 9–10 × 6 µm. *Poisonous.*

5 Inocybe obscuroides: *pale cigar-brown rather scaly cap with entirely lilac-violet, short thick stem*

Leaf-litter in mixed woods. Autumn. Uncommon. Cap 2–4 cm (1–1½ in), convex then expanded with low umbo, occasionally pale lilac at margin; woolly-fibrillose with minute, dark brown concentric scales, erect at centre, more fibrillose at margin. Gills whitish with lilac tint, soon cigar-brown. Stem not bulbous, lacking fibrils or scales, paler buff at base. Spores dull brown, elliptic, 8–10 × 5–6 µm. Odour strong, pungent. *Poisonous.* The related *I. griseolilacina* (pp. 94–5) has a longer, slender stem and paler evenly scaled cap; *I. cincinnata* has violet at stem apex only; both have fibrillose-scaly, not smooth, stems.

HEBELOMA

1 *Hebeloma mesophaeum: in heathy woodlands near birch; date-brown cap with distinct paler margin; ring-like veil remnants on stem*

Autumn. Common. Cap 2–5 cm (1–2 in), convex then flattened, slightly viscid. Gills sinuate, dull ochre-brown. Stem whitish to brown. Flesh dark brown in stem base. Spores elliptic, 9–11 × 4–5 μm. Edibility doubtful; best avoided; has bitter taste.

2 *Hebeloma radicosum: stem deeply "rooting"; laurel-marzipan smell*

Often associated with rodent and kingfisher nests. Autumn. Uncommon. Cap 5–10 cm (2–4 in), convex then umbonate, clay-brown; smooth and slightly viscid with faint, slightly darker adpressed scales scattered on surface. Gills sinuate, whitish then clay-brown. Stem stout and bulbous when young but soon elongating, difficult to pick due to the "rooting" base; scaly bands or belts below the distinct ring zone. Spores dull brown, elliptic, 8–9 × 5 μm. Odour especially strong in gills when bruised. Edibility doubtful; best avoided.

3 *Hebeloma crustuliniforme* (Poison Pie): *pale buff coloration; stout stem; often strong smell of radish*

In mixed woods, gardens and hedgerows. Autumn. Common. Cap 3–6 cm (1–2½ in), convex then expanded with low umbo, margin inrolled when young; cuticle smooth and slightly viscid when moist. Gills sinuate, whitish then dull clay-brown, exuding droplets in wet weather. Stem variable but usually rather short, stout, white with "powdery" apex. Spores cigar-brown, almond-shaped, 10–12 × 5–7 μm. *Poisonous.*

4 *Hebeloma sinuosum: large pale-coloured species; stem with scaly zones*

Mixed woods, hedgerows etc. Autumn. Uncommon. Cap 5–15 cm (2–6 in), convex then soon flattened, smooth and slightly viscid; pale tan to fleshy-buff. Gills sinuate, narrow, whitish then dull brown. Stem tall, stout, slightly rooting; whitish with scaly zones. Spores ovate, 10–12 × 5–7 μm. *Poisonous.*

5 *Hebeloma testaceum: brick-red to tan cap; slight veil remnants on upper stem*

Mixed woods and heaths. Autumn. Uncommon. Cap 3–5 cm (1–2 in), convex then flattened; brick-red to tan, paler at margin; smooth and slightly viscid. Gills sinuate, whitish then clay-brown to cigar-brown. Stem white, darker reddish below. Spores ovate, 8–9 × 4–5 μm. Edibility doubtful; best avoided.

CORTINARIUS

All members of *Cortinarius* have rust-brown spores, but the genus is often divided into groups (subgenera) according to certain features: *Myxacium*, medium to large species with viscid caps and stems; *Phlegmacium*, large fleshy species with viscid or moist caps but dry stems; *Cortinarius*, medium to large, with dry, scaly to moist caps, and dry stems, sometimes robust; *Dermocybe*, medium-sized, rarely large, caps dry, silky or fibrillose, often yellow-orange to reddish, stems slender and dry; *Sericeocybe*, small to medium-sized, caps silky-fibrillose, not hygrophanous, sometimes sticky in wet weather and then often robust, stem bulbous; and *Hydrocybe*, small, distinctly hydrophanous (changing colour with moisture).

1 Cortinarius *(Myxacium)* **delibutus:** *clear yellow cap and stem, both viscid; young gills and stem apex bluish-violet*

Under birch and aspen on boggy soil. Autumn. Occasional. Cap 5–10 cm (2–4 in), convex then expanded, smooth, viscid; clear golden to brownish yellow. Gills bluish lilac when young then soon rust-brown. Stem tall, clavate; viscid, with reddish ring zone; pale yellow below. Spores subglobose, 7–9 × 6–7 μm. Not recommended, taste bitter.

2 Cortinarius *(Myxacium)* **pseudosalor** *(=C. elatior in part): shiny, dull yellow-ochre to brown cap with uneven radial wrinkles; slimy violet-tinted tapered stem*

Under beeches especially on clay soils. Summer and autumn. Common. Cap 3–10 cm (1–4 in), conical then expanded. Gills broad, adnate, violaceous then soon rust-brown, strongly interconnected with cross-veins. Stem cylindrical, tapering, slightly rooting; white, viscid and often strongly tinted violet below with belt-like zones. Spores warty-rough, rust-brown, 12–15 × 7–8 μm. Edible but not highly recommended.

3 Cortinarius *(Myxacium)* **ochroleucus:** *pale cream cap and stem; only very slightly viscid; taste bitter*

In beech woods on clay soils. Autumn. Uncommon. Cap 3–8 cm (1–3 in), convex with a low umbo, pale cream to ochre; smooth, silky-fibrillose to slightly viscid. Gills pale yellow-ochre then rust-brown. Stem tapering, almost white, slightly viscid becoming dry and silky. Spores 7–8 × 4–5 μm. Not recommended for eating.

CORTINARIUS (contd)

1 *Cortinarius* (Phlegmacium) auroturbinatus: *bright golden colours; strong aromatic smell; stem with large flattened basal bulb*

Under beeches on chalky soil. Autumn. Occasional to frequent. Cap 4–10 cm (1½–4 in), convex then expanded; viscid, bright golden-orange. Gills yellow when young then soon rust-brown. Stem rather tall, stout, with bulb; fibrillose, sulphur-yellow becoming paler above. Flesh sulphur-yellow in stem base, almost white above and in cap. Spores lemon-shaped, 12–15 × 7–8 μm. Odour strong, aromatic. Edible but not recommended.

2 *Cortinarius* (Phlegmacium) glaucopus: *smooth, viscid, rusty-brown cap (greenish when young); stem apex and young gills bluish; marginate bulb*

In mixed woods, sometimes in large numbers. Autumn. Frequent. Cap 6–12 cm (2½–4½ in), convex then soon flattening; fleshy, edge undulate; smooth, slightly viscid, greenish to ochre then rust-brown, but variable. Gills crowded, emarginate, azure-blue then soon brownish. Stem rather short, swollen at base with marginate bulb; ochre or brownish below, lilac-blue at apex; fibrillose with clinging veil remnants, often stained rust-brown with spores. Spores elliptic, 9–10 × 5–6 μm. Edible but of low quality.

3 *Cortinarius* (Cortinarius) alboviolaceus: *under beech and oak; pale lilac-white colours*

Autumn. Frequent to common. Cap 3–8 cm (1–3 in), convex then expanded and umbonate. Silky-fibrillose; pale bluish-lilac when fresh, fading to white or greyish with age. Gills greyish lilac then rust-brown, rather broad, distant. Stem swollen, clavate, bluish-lilac above, paler below reaching up to a cobweb-like veil. Flesh bluish, especially in stem. Spores ovate, 7–9 × 6–7 μm. Edible but not highly recommended.

CORTINARIUS (contd)

1 *Cortinarius* (*Phlegmacium*) multiformis: *buff-yellow cap; stem with basal bulb; flesh smells acidic, of apples*

In mixed woods. Autumn. Occasional. Cap 5–10 cm (2–4 in), convex then expanded and slightly umbonate, ochre to buff with yellower margin, slightly viscid. Gills pale whitish-clay at first then rust-brown. Stem cylindrical with prominent basal bulb (usually marginate, sometimes merely rounded), pale yellowish. Flesh soft, yellowish. Spores 10–11 × 5–7 µm. Edible and good.

2 *Cortinarius* (*Hydrocybe*) armillatus (Red-banded Cortinarius): *brick-red cap; pale stem with red belts*

Under birch in bracken and heath vegetation. Summer and autumn. Frequent to common. Cap 5–12 cm (2–4½ in), convex then expanded and flattened; dry, fibrillose, brick-red to tawny-orange when old. Gills whitish then cinnamon-brown. Stem tall, swollen-clavate; pale whitish-brown with three or four reddish belts running obliquely round. Spores almond-shaped, 9–12 × 5–6 µm. Edible but not recommended.

3 *Cortinarius* (*Phlegmacium*) purpurascens: *brownish cap with bluish-violet stem; flesh bruising deep violet*

In mixed woods. Autumn. Common. Cap 6–15 cm (2½–6 in), convex then expanded and flattened; yellowish to dark umber brown, occasionally with violet tint at margin, slightly viscid. Gills deep violet at first then cinnamon- to rust-brown. Stem rather stout, bulbous, often with a clearly marginate bulb but variable; fibrillose, pale bluish violet bruising deeper violet; cortinal zone at apex. Spores 8–11 × 5–7 µm. Edible but not highly recommended.

4 *Cortinarius* (*Cortinarius*) violaceus: *entire fruit-body an intense deep violet; velvety-floccose cap*

In mixed woodlands, especially birch. Autumn. Uncommon. Cap 5–15 cm (2–6 in), convex then expanded; dry, velvety, deep and intense violet to violet-brown. Gills broad, distant, violet then rust-brown. Stem tall, fibrillose, swollen at base, also violet. Spores 11–14 × 7–9 µm. Odour distinctive of cedar. Edible and good.

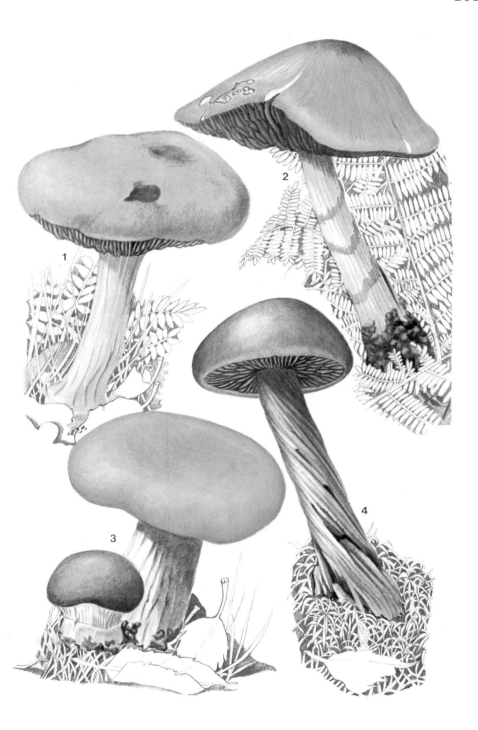

CORTINARIUS (contd)

1 *Cortinarius* (Dermocybe) anomalus: *gills and stem apex violet when young, soon fading to an even cinnamon-brown in the gills, pale yellowish in stem*

In leaf-litter of birch, beech or oaks. Autumn. Common. Cap 3–7 cm (1–3 in), convex-umbonate then soon flattened; smooth or very slightly fibrillose; dull clay or reddish brown, with a slight violet tint when young. Gills at first deep violet then soon cinnamon brown. Stem long, clavate; violet at apex when young, pale yellowish cream below; surface slightly squamulose to almost smooth. Spores ovate, 7–9 × 6–7 μm. Edible but not highly recommended.

2 *Cortinarius* (Sericeocybe) traganus: *stout lilac-blue fungus, discolouring patchily to ochre; strong goaty smell*

In northern conifer woods. Autumn. Occasional to frequent. Cap 5–10 cm (2–4 in), convex to broadly umbonate, thick-fleshed, dry, smooth, silky; often cracking. Gills broad, adnexed, widely spaced; ochre-yellow then rust-cinnamon. Stem stout, clavate-bulbous, lilac-blue to violet with remains of woolly veil around base and middle. Flesh in cap ochre-yellow becoming darker in stem to rust-brown at base. Spores elliptic, 8–10 × 5–6 μm. Inedible.

3 *Cortinarius* (Dermocybe) orellanus: *rough, reddish or pinky-brown cap and often spindle-shaped yellow stem; gills widely spaced*

In mixed woodlands, especially birch and oaks. Autumn. Uncommon. Cap 3–5 cm (1–2 in), convex, and often umbonate, dry, fibrillose, with slightly scurfy-scaly surface; the minute scales are a darker more olive brown than the rest of the cap. Gills widely spaced, broad, yellowish orange then soon cinnamon-brown. Stem often spindle-shaped, occasionally clavate; yellowish, fibrillose. Cobweb-like cortina visible between cap and stem when young. Spores subglobose, 8–11 × 5–8 μm. *Highly poisonous*, even *fatal*, apparently often many days after eating the fungus.

4 *Cortinarius* (Dermocybe) uliginosus: *cap sharply conical-umbonate, intense orange-red; in sphagnum moss under aspens and willows, in boggy situations*

Autumn. Uncommon. Cap 2–5 cm (1–2 in), dry, silky-fibrillose; browner with age. Gills yellow then soon olive and finally cinnamon-brown. Stem curved, same colour as cap, paler above, with cobweb-like veil. Spores elliptic, 8–9 × 4–5 μm. Edibility doubtful; best avoided. One of the most attractively coloured of the smaller cortinarii.

CORTINARIUS (contd)

1 *Cortinarius* (Hydrocybe) *hinnuleus:* umbonate ochre cap; stem white-zoned; strong smell of coal gas especially in old specimens

In grass under deciduous trees, especially oak. Autumn. Occasional to common. Cap 3–6 cm (1–2½ in), convex with broad umbo; smooth, dry, often radially wrinkled; yellow-ochre to dull brown, paler at edge. Gills ochre then soon rust-brown; broad and rather distant. Stem equal or tapering downward, pale whitish ochre, a distinct white belt-like zone around middle and a cobweb-like cortina above this. Spores ovate, 6–9 × 4–5 μm. Edibility very doubtful; best avoided.

2 *Cortinarius* (Sericeocybe) *pholideus:* hazel-brown scaly cap and stem; stem with lilac tints when young

In leaf-litter under birches. Autumn. Frequent. Cap 3–8 cm (1–3 in), convex with low umbo, covered with minute recurved, pointed scales, darker than surface underneath. Gills crowded, broad, lilac at first then soon cinnamon-brown. Stem tall, rather slender, lower half with scales like cap, upper half at first tinged lilac from the cortina. Spores elliptical, 6–7 × 4–5 μm. Edibility doubtful; best avoided.

GYMNOPILUS

3 *Gymnopilus penetrans:* small yellow-orange caps and stems; gills spotted rust-brown

On twigs, chips or logs of birches and conifers. Summer and

autumn. Common or even abundant. Cap 1–5 cm ($\frac{1}{2}$–2 in), convex then expanded; dry, slightly scaly-fibrillose to almost smooth; golden tawny with paler veil remnants at margin. Gills thin and crowded, golden yellow but soon with rust-coloured stains. Stem rather short, slender, yellow-orange, paler above and usually with white base; very faint ring zone of veil at apex. Spores rich orange, elliptic, slightly warty, 7–8 × 4–5 μm. Inedible, usually very bitter.

4 *Gymnopilus junonius* (=Pholiota spectabilis): *large orange cap and stem with thick ring; in clumps always at base or on stumps, logs of deciduous trees*

Summer and autumn. Common. Cap 5–15 cm (2–6 in), convex then expanded; dry, fibrillose, sometimes slightly scaly; rich bright orange. Gills thin, crowded, yellow then rust-orange. Stem large, swollen at base, fibrillose-scaly, orange with thick fleshy ring at apex. Spores ovate, rust-orange, 8–10 × 5–6 μm. Inedible and possibly hallucinogenic; flesh extremely tough and fibrous as well as rather bitter. This is one of the commonest and most spectacular of the fungi found on wood.

CREPIDOTUS

1 ***Crepidotus mollis:*** *gelatinous shell- or kidney-shaped cap with tiny lateral stem*

On stumps and logs of deciduous trees, especially oaks. Summer and autumn. Common. Cap 2–7 cm (1–3 in), pale ochre to almost white when dry, with a thick gelatinous cuticle. Gills crowded, thin, clay-brown to cinnamon. Stem short, almost absent, lateral. The whole fungus is horizontal, bracket-like. Spores clay-brown, smooth, 6–8 × 4–6 μm. Edibility uncertain; best avoided.

GALERINA

2 ***Galerina (=Pholiota)*** ***mutabilis:*** *cap changes colour as it dries, often bicoloured; stem scaly; in clumps*

On stumps of deciduous trees. Summer and autumn. Common. Cap 3–8 cm (1–3 in), convex then flattened with a low umbo, smooth; rich date-brown when wet then drying out a pale chamois-leather colour, usually with the umbo retaining the darker, moist colouring long after the rest has dried; may have a darker margin. Gills adnate, pale ochre then cinnamon-brown. Stem dark brown and scaly below the ring, paler and smooth above. Ring membranous, flaring, usually rust-brown from spore deposit. Spores ovate, 6–8 × 4–5 μm. Edible and excellent, but note that this species is found *in clumps* – similar but solitary species can be poisonous.

3 ***Galerina hypnorum:*** *caps very small, striate almost to the centre; pale orange colours; always in moss*

In moss on logs, stones, soil etc. Spring to autumn. Common. Cap 4–6 mm (¼ in), hemispherical, yellow-brown to orange. Gills broad, distant, adnate, pale tawny-yellow then rust-brown. Stem slender, smooth, pale yellow. Spores rust-brown, slightly warty, 10–11 × 6–7 μm. Edibility uncertain but too small to be worth considering. There are a few very similar species, also found in mosses, which are difficult to separate without a microscope.

PHAEOLEPIOTA

4 ***Phaeolepiota (=Pholiota)*** ***aurea:*** *large scurfy-granular cap; tall stem also scurfy-granular up to large ring*

On bare soil in mixed woodlands, often in large troops. Autumn. Rare to occasional. Cap 5–20 cm (2–8 in), convex then flattened with low umbo, tawny to golden-brown, with a granular, scurfy surface. Gills adnexed, pale ochre-yellow then rust-coloured. Stem tall, with a sheath- or stocking-like granular veil covering stem and flaring outwards to form the large ring, tawny-brown; stem apex smooth, paler. Spores ochre, smooth, warty, 11–14 × 4–5 μm. Edible but poor; not recommended.

ENTOLOMATACEAE

The fungi of this large group all have a pink coloration to the spore deposit. The spores are unusual in being very angular and polygonal, rather like some crystals in geological deposits. Gill attachment varies between genera from sinuate or emarginate to adnate. Some colours not usually found within the fungi (blue, violet, green) are often found in this attractive family. Genera included here are: *Entoloma, Leptonia*. None is recommended for the table.

ENTOLOMA

1 *Entoloma sinuatum* (=*lividum*) (Livid Entoloma): *in woodland clearings; pale wavy-edged cap silky; gills yellowish then pink*

Similar in many respects to the more common *E. clypeatum* illustrated below, but found in open spaces in deciduous woodlands in early autumn and rather uncommon. It differs in the paler grey to fawn cap, which is smooth and evenly coloured; margin at first inrolled and then expanded and wavy. Stem longer than in *E. clypeatum* but still rather stout, white. Spores 9–11 × 8–9 μm. *Poisonous*, causing severe sickness and diarrhoea, but rarely fatal.

2 *Entoloma porphyrophaeum*: *tall, fibrous, with wine-purple tints in the brown cap and stem*

In fields and woodland clearings. Summer and autumn. Occasional. Cap 3–8 cm (1–3 in), campanulate then expanded, umbonate; fibrillose, brown with distinct wine-coloured tint. Gills sinuate to emarginate (notched near stem), whitish brown then flesh-pink. Stem tall, fibrous, often twisted, purplish brown with white base. Spores 10–13 × 6–7 μm. Apparently edible but not recommended in case of confusion.

3 *Entoloma clypeatum*: *under fruit or rose bushes; irregularly shaped and unevenly coloured cap; smell floury, of new meal or of cucumber*

Spring, summer and early autumn. Occasional to frequent. Cap 3–7 cm (1–3 in), expanded with an obtuse umbo, edge of cap often wavy, irregular; olive-grey to yellowish brown with darker, irregular streaks. Gills sinuate, pale grey then soon flesh-pink. Stem short, stout, white, flesh fibrous. Spores 9–12 × 7–9 μm. *To be avoided* as there are similar and very dangerous poisonous species.

4 *Entoloma sericatum*: *thin fragile cap and slender stem; strong smell of bleach*

In mixed woodlands in rather damp, boggy places. Autumn. Frequent. Cap 2–4 cm (1–1½ in), umbonate then flattened; fragile, smooth and hygrophanous; pale fawn to brown. Gills adnate, pale greyish yellow then pink. Stem long, thin, fragile, almost white. Spores 8–10 × 6–8 μm. *Poisonous*.

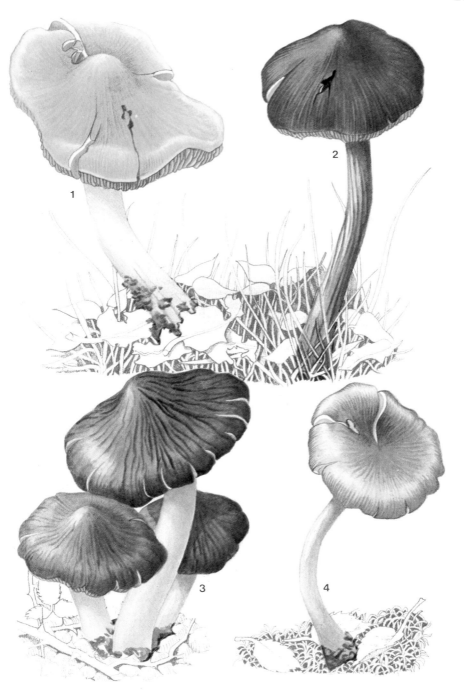

ENTOLOMA (contd)

1 *Entoloma rhodopolium:* grey-brown cap drying much paler; white stem; smells and tastes like new meal

In grass and leaf-litter in deciduous woods. Summer and autumn. Common. Cap 3–7 cm (1–3 in), expanded and umbonate then flattened, margin slightly inrolled; greyish brown to yellowish brown but paler, ash-grey and silky when dry. Gills emarginate, white then flesh-pink. Stem slender, cylindrical, white to pale greyish, fibrous, fragile. Spores 8–10 × 6–8 μm. *Poisonous* although not deadly; causes gastric upsets.

2 *Entoloma nitidum:* both conical-umbonate cap and stem deep blue; in damp woods

On boggy soil under birch and pines. Autumn. Uncommon. Cap 2–5 cm (1–2 in), conical then expanded and umbonate; fibrillose to minutely scaly at centre; deep indigo blue. Gills broad, emarginate, white then pink. Stem long, tapering and "rooting", fibrous; blue as cap but paler, whitish at base. Spores 7–9 × 6–8 μm. Probably not poisonous but best avoided, as are all toadstools in this family.

LEPTONIA

3 *Leptonia chalybaea:* convex cap and stem, both deep blue-black to violet; gills pale grey-blue then pink

In open fields and pastures. Summer and autumn. Occasional to frequent. Cap 2–3 cm (1 in), convex to slightly umbonate; dry and minutely squamulose; intense blue-black to violet-blue. Gills broad, crowded, adnate. Stem long, slender, cartilaginous, dark blue. Spores 9–10 × 7–8 μm. Edibility uncertain; best avoided.

TRICHOLOMATACEAE

Most of this very large family have a white to cream or pale pinkish spore print but they differ markedly in appearance, habitat, and chemical make-up. For example *Tricholoma* consists entirely of terrestrial, rather fleshy, fibrous species with sinuate gills and white spores, while *Panellus* is found on trees and logs with almost stemless brackets, with spores pinkish clay in deposit. These differences are reconciled by similarities in development and anatomy.

CLITOPILUS

1 Clitopilus prunulus (The Miller): *soft white cap and stem with "kid glove" texture; pink spores; mealy smell*

On soil in mixed woodlands. Autumn. Common. Cap 2–8 cm (1–3 in), convex then expanded and funnel-shaped, margin irregular, wavy; white to pale cream, smooth. Gills more or less decurrent, white then pinkish. Stem from very short to long, white, fibrous. Spores fusiform with longitudinal ribs, pink, 10–14 × 4–6 μm. Edible and delicious.

TRICHOLOMA

2 Tricholoma flavovirens (=equestre): *greenish yellow cap, centre reddish; gills, stem sulphur-yellow; under conifers*

Autumn. Frequent. Cap 5–12 cm (2–4 in), convex then expanded with low umbo; smooth, slightly viscid; sulphur to olive-yellow; minutely scaly, reddish, towards centre. Gills rather crowded, sinuate-emarginate. Stem stout, fibrillose, paler. Flesh with slight mealy smell. Spores white, 5–8 × 4–5 μm. Edible and recommended. Note that the Death Cap (*Amanita phalloides*) which sometimes has a similarly coloured cap, is much softer-fleshed and has both ring and volva.

3 Tricholoma columbetta: *pure almost dazzling silky white, sometimes with small blue stains when old; no odour*

In leaf-litter of beech woods. Autumn. Occasional. Cap 6–10 cm (2½–4 in), convex then expanded with slightly fibrillose margin. Gills crowded, sinuate, white. Stem tall, fibrous, rooting, pure white. Spores white, 5–7 × 4–5 μm. Edible and recommended.

4 Tricholoma albobrunneum: *slightly viscid chestnut-brown cap; often bicoloured stem; under conifers*

Autumn. Occasional to frequent. Cap 5–10 cm (2–4 in), convex then expanded, margin incurved, often scalloped; smooth and slightly viscid, finely fibrillose, chestnut-brown. Gills sinuate-emarginate, white, rather crowded. Stem stout, cylindrical, fibrous; white above, brownish below, often clearly defined like a band. Spores white, elliptic, 4–6 × 4 μm. Edible, but considered of low quality, even disagreeable.

TRICHOLOMA (contd)

1 *Tricholoma ustaloides:* viscid brown cap; white gills spotted brown; floury smell and taste

In mixed woods. Autumn. Occasional to frequent. Cap 3–10 cm (1–4 in), convex then expanded with margin often wavy and irregular and incurved when young; viscid, rich chestnut to bay-brown. Gills white, sinuate, becoming spotted with brown stains. Stem slender to medium thickness with bulbous base, fibrillose; white above and pruinose becoming darker brownish at base. Flesh reddening slightly. Spores white, 5–7 × 3–5 μm. Edibility: not recommended.

The very similar *T. ustale* found in the same habitat differs mainly in the flesh not smelling of new meal and the redder cap.

2 *Tricholoma fulvum:* reddish brown streaky cap and stem; yellow stem flesh

Usually under birches in peaty, boggy soil. Autumn. Common. Cap 4–10 cm (1½–4 in), convex and rather umbonate, then expanded; fibrillose or with radial streaks, viscid in damp weather; reddish brown. Gills sinuate, yellowish with brown spotting. Stem rather long, fibrillose, reddish brown. Flesh in stem yellow, in cap whitish, with mealy odour and taste. Spores white, 5–7 × 3–5 μm. Not recommended although not poisonous.

3 *Tricholoma orirubens:* dark, almost black cap, scaly-squamulose; pink stains often (not always) on gills; strong floury smell and taste; yellow mycelium

In coniferous woodlands. Autumn. Occasional to common. Cap 2–7 cm (1–3 in), convex then expanded usually with an umbo; deep grey to black with numerous fine scales. Gills sinuate, white often with black edging and frequently turning rose-pink where eaten by insects etc. or otherwise damaged. Stem smooth and white or sometimes with grey punctae. Spores white, 4–6(7) × 3–4 μm. Edible and delicious.

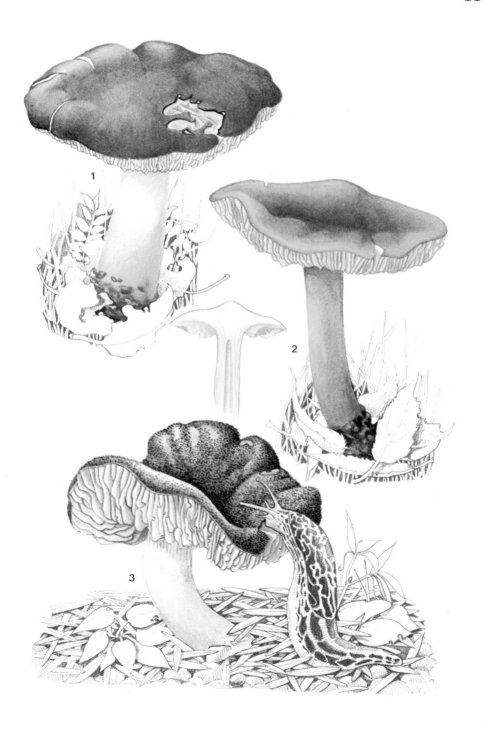

TRICHOLOMA (contd)

1 *Tricholoma pardinum: coarsely scaled grey-brown cap, often large; gills with olive tints, exuding droplets in damp weather*

In conifer woods chiefly in mountainous regions of Europe, not yet found in Britain. Autumn. Occasional. Cap 9–20 cm (3½–8 in), convex then expanded and broadly umbonate, margin inrolled; usually with rather large, tomentose scales darker grey-brown on a pale brown or grey background. Gills sinuate, white to slightly yellowish, finally with olivaceous tints. Stem stout, fleshy, fibrous, with a white tomentose surface above becoming slightly darker, brownish, below. Flesh smells slightly of flour. Spores white, large, 8–10 × 5–6 μm. *Dangerously poisonous* although not usually deadly, causing severe gastroenteritis.

2 *Tricholoma portentosum: dark smooth cap with very fine fibrils; yellowish gills and stem; stout build; no distinctive smell*

Often below conifers but sometimes under mature beeches. Autumn. Occasional. Cap 5–10 cm (2–4 in), convex then soon expanded and often umbonate; smooth with fine radiating fibrils; dark greyish to greenish brown, darker almost black at centre, where the fibrils go black. Gills white or yellowish, sinuate. Stem stout, smooth white or tinged yellow. Spores white, 5–6 × 3–5 μm. Edible and delicious.

3 *Tricholoma psammopum: only under larch; dry, rough, reddish brown cap and densely punctate stem*

Autumn. Occasional to frequent. Cap 3–8 cm (1–3 in), convex then expanded and umbonate; dry, minutely roughened; reddish brown to tan. Gills pale yellow-brown with brownish stains. Stem rather slender; white at apex, brownish below, with dark brown dense punctae. Spores 5–6 × 4–5 μm. Edible but not recommended.

4 *Tricholoma imbricatum: reddish brown, umbonate, almost conical cap with overlapping scales*

Always below pines. Autumn. Frequent to common. Cap 4–10 cm (1½–4 in), convex to conical, then expanded with broad umbo; dry, dull reddish brown to umber, surface cracking all over into rough overlapping (imbricate) scales. Gills sinuate, white then pale brownish, spotted darker brown. Stem rather stout, swelling at base, pale brown, fibrous; often rather hollow. Spores white, 5–7 × 3–5 μm. Edible but not highly recommended.

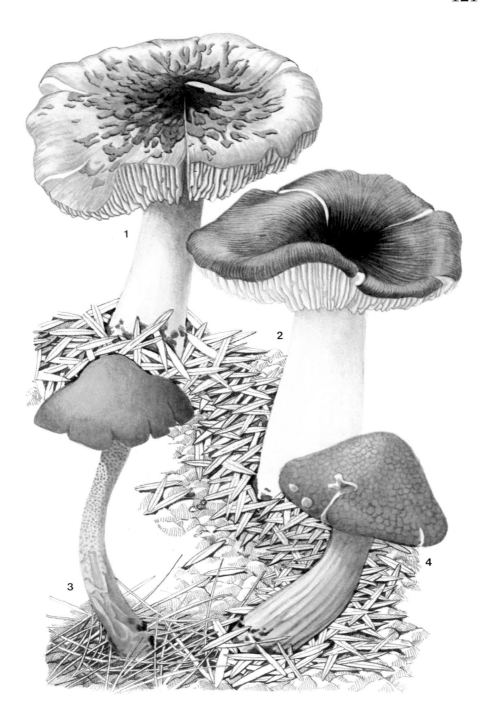

TRICHOLOMA (contd)

1 *Tricholoma saponaceum* (Soap-scented Toadstool): *cap colours variable, often pale but normally blackish at centre; pinkish flush in stem and gills; distinctive smell of cheap soap*

In mixed woods on soil, often in small clumps. Summer and autumn. Common. Cap 3–10 cm (1–4 in), convex then expanded and broadly umbonate; smooth, dry or very slightly squamulose at centre; colour very variable, even white, to olivaceous or brown, usually blackish at centre. Gills sinuate, rather distant; white or tinged sulphur-yellow or greenish, often speckled with reddish spots. Stem rather stout, tapering and rooting, fibrous; white or same colour as cap, often slightly squamulose. Spores white, 5–6 × 3–4 µm. Not poisonous but of very poor quality and rather bitter.

2 *Tricholoma sejunctum*: *yellow-green fibrillose cap; white gills and stem flushed yellow; floury smell*

Mainly deciduous woodlands. Autumn. Occasional to frequent. Cap 5–10 cm (2–4 in), convex then expanded and often sharply umbonate; pale to greenish yellow with darker radiating fibrils and streaks; smooth or slightly squamulose, viscid in wet weather. Gills rather distant, broad, strongly emarginate; white or with yellow tints. Stem fibrous, tough, often rather wavy and irregular in shape; white with pale yellowish tints. Taste also floury but then bitter. Spores white, 5–7 × 4–5 µm. Not edible; has unpleasant taste and can cause upsets.

3 *Tricholoma sulphureum* (Sulphur Toadstool): *strong sulphur-yellow in all parts; strong specific smell of coal gas*

Usually under oaks. Autumn. Occasional to frequent. Cap 3–8 cm (1–3 in), convex then expanded and broadly umbonate; smooth, dry; clear sulphur-yellow to slightly brownish at centre. Gills distant, broad and thick, sinuate-emarginate, sulphur-yellow. Stem rather long, fibrous and tapering, sulphur-yellow. Spores white, 8–11 × 5–6 µm. Not edible.

4 *Tricholoma argyraceum*: *greyish, minutely squamulose cap; often yellowish tints in gills and stem; floury smell and taste*

Usually under beech but also conifers. Autumn. Frequent to common. Cap 4–8 cm (1½–3 in), convex then expanded and umbonate; brownish grey and squamulose-fibrillose all over; edge with remains of fine veil. Gills rather crowded, sinuate; white or greyish, ageing yellowish. Stem white, yellowish or greyish, smooth, rather slender. Spores white, 5–6 × 3–4 µm. Edible.

The closely related *T. terreum* (mixed woods, common, edible) differs by its rather darker blackish grey cap, greyish gills, lack of the floury odour and taste, and larger spores (6–7 × 4–5 µm).

1
2
3
4

TRICHOLOMA (contd)

1 Tricholoma cingulatum: *small, usually under willows; persistent cottony ring; slightly floury smell and taste*

Autumn. Uncommon. Cap 2–6 cm (1–2½ in), convex then expanded and slightly umbonate; brownish grey and minutely velvety-squamulose. Gills distant, sinuate, white. Stem rather long, slender, white and slightly fibrillose; distinct and persistent cottony ring near apex. Spores white, 4–6 × 2–4 µm. Edible.

CALOCYBE

2 Calocybe gambosa *(=Tricholoma gambosum)* (St. George's Mushroom): *only in spring and early summer; all parts creamy-white; strong floury taste and smell*

In fields, woodland margins and roadsides. Frequent to common. Cap 4–12 cm (1½–4½ in), convex then expanded with wavy, irregular margin; smooth, never scaly or fibrillose; ivory-white to pale buff. Gills sinuate to slightly decurrent when cap expanded, very crowded; white to cream. Stem short and stout, flesh fibrous; ivory-white. The floury taste and smell disappear on cooking. Spores white, 5–6 × 3–4 µm. Edible and delicious. This is one of the best edible species and cannot easily be confused except perhaps with *Inocybe patouillardii* (pp. 92–3) which can appear very early and bears a *very* superficial resemblance.

LYOPHYLLUM

3 Lyophyllum decastes *(Tricholoma aggregatum): in dense clusters with stem bases fused into one common base; grey to brown colour*

In woods, on pathsides. Autumn. Common. Cap 4–10 cm (1½–4 in), convex then expanded, margin incurved and usually wavy; smooth, dry, minutely fibrillose; grey-brown to yellow-brown. Gills crowded, adnate to decurrent, white to greyish. Stem tough, fibrous, many joining together at the base; white or greyish, floccose-pruinose at apex. Spores white, globular, 6 × 5–6 µm. Edible but not highly recommended.

4 Lyophyllum connatum *(=Clitocybe connata): in smallish clusters, bases of stems fused but distinct; chalk-white colour; specific chemical test*

In clumps (usually smaller than the previous species). In grassy places in woodlands and pathsides. Autumn. Uncommon. Cap 4–10 cm (1½–4 in), convex then expanded, margin incurved and wavy; smooth, dry; pure chalk-white. Gills crowded, narrow, adnate-decurrent, white to yellowish. Stems white, long and slender, not springing from one common base. Spores white, elliptic, 6–7 × 3–4 µm. Chemical test: ferrous sulphate on gills = violet in one minute. Edible but not recommended – easily confused with poisonous species of *Clitocybe*.

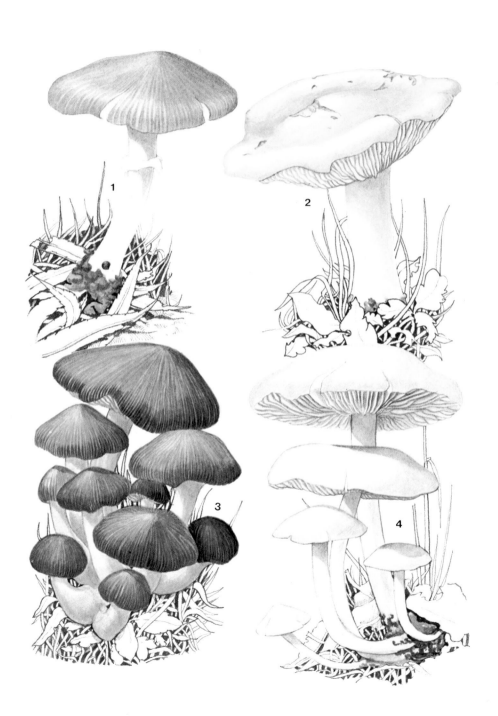

1 *Melanoleuca grammopodia: very large umbonate cap; rather pale brown colours; often in large rings*

In woodland clearings, meadows etc. often in large rings, especially in mountainous districts. Autumn. Occasional. Cap 6–25 cm (2½–10 in), convex then expanded and umbonate, grey-brown to yellow-brown, hygrophanous, umbo usually darker. Gills crowded, sinuate-adnate, white then pale cream. Stem tall, rather stout, straight, slightly bulbous; whitish with brownish fibrils. Flesh white to pale brown, strong, rather unpleasant smell. Spores white, elliptic with warts, amyloid (turning blue-black) with iodine, 8–10 × 5–6 µm. Best avoided although not recorded as poisonous.

2 *Melanoleuca cognata: rather yellowish or tan coloration; stem tall; gills ochre-tan*

In conifer woods on paths and clearings. Autumn. Uncommon. Cap 5–10 cm (2–4 in), convex then expanded and umbonate; smooth, ochre-yellow to tan, paler when dry. Gills becoming clearly pale ochre or tan when mature; sinuate-adnate, crowded. Stem tall, straight, slightly bulbous, colour as cap with darker fibrils. Flesh white to pale ochre, with pleasant odour. Spores white, elliptic, with amyloid warts, 7–9 × 5–6 µm. Edible and quite tasty.

3 *Melanoleuca evenosa: very pale, almost white coloration; in spring in grass*

In pastures, meadows, woodland paths. Spring. Frequent. Cap 4–8 cm (1½–3 in), convex then expanded and umbonate; pale ivory to greyish, hygrophanous, becoming white on drying. Gills whitish or with very pale flesh tints; crowded, sinuate-adnate. Stem slender, straight; white with slightly darker fibrils. Flesh white. Odour rather strong, fragrant. Spores white, elliptic, with amyloid warts, 8–9 × 5–6 µm. Edible but not highly recommended.

4 *Melanoleuca melaleuca (=vulgaris): cap umbonate, deep brown but hygrophanous; stem fibrillose; gills white*

In mixed woods, especially pathsides. Autumn. Common. Cap 4–10 cm (1½–4 in), convex then expanding and usually slightly umbonate; smooth; deep brown when moist, drying almost buff. Gills broad, crowded, sinuate-adnate, white. Stem rather tall, straight, slightly swollen at base; whitish (when dry) to brown, with darker, brown fibrils. Flesh white then pale brownish in stem. Odour pleasant. Spores pale cream, elliptic, with minute amyloid warts, 8 × 4–5 µm. Edible but not recommended. Like all *Melanoleuca* species this has characteristic microscopic harpoon-shaped sterile cells (*cystidia*) among the basidia on the gill surface.

LEPISTA

1 *Lepista nuda (=Tricholoma nudum)* (Wood Blewits): *lilac or violet colours in cap, gills and stem; in woods and gardens in leaf-mould*

Late autumn. Common. Cap 4–10 cm (1½–4 in), convex then expanded, flattened; smooth, greasy; lilac to reddish violet, browner with age. Gills crowded, adnate then slightly decurrent; soft, lilac, easily separable from cap tissue. Stem rather short, stoutish, fibrous, lilac-violet. Spores flesh-pink in a deposit, 6–8 × 4–5 μm. Edible and delicious; one of the very best species, but one to which a few people are allergic: try only a little the first time.

2 *Lepista saeva (Tricholoma personatum)* (Common Field Blewit, Blue-leg): *cap and gills without lilac colours; stem bluish-lilac; in grassy open places*

Late autumn, early winter. Frequent to common. Cap 4–10 cm (1½–4 in), convex then expanded, margin inrolled; smooth, dry, pale buff or clay to greyish. Gills crowded, adnate-decurrent; white or pinkish-buff, never lilac. Stem short, stout, fibrillose, clear bluish-lilac. Spores flesh-pink, 5–8 × 3–5 μm. Edible and delicious; one of the best and best-known species, but one to which a few people are allergic.

3 *Lepista irina (=Tricholoma irinum): cap, gills and stem lacking lilac tints; sweet smell of violets or mock orange; in clearings and at woodland margins*

Late autumn. Occasional to frequent. Extremely similar to the two previous species, but completely lacking lilac or violet tints, having a tan cap, whitish stem and white-clay gills. Spores flesh-pink, 7–8 × 3–4 μm. Edible and delicious.

NYCTALIS

4 *Nyctalis (=Asterophora) parasitica: only on old fruit-bodies of* Russula *and* Lactarius

Autumn. Occasional. Cap 1–3 cm (½–1 in), convex then flattened; smooth, silky, grey; slightly striate at margin. Gills thick, distant, whitish then soon covered in a brownish powder-like deposit which is composed of special spores called chlamydospores (the similar *N. lycoperdoides* turns powdery *all over*). Stem slender, white, smooth. Smell repulsive, pungent. Spores buff, 5–6 × 3–4 μm, but usually replaced by the asexually produced chlamydospores, 15 × 10 μm. Not edible.

CLITOCYBE

1 *Clitocybe nebularis* (Clouded Agaric): *in rings; umbonate, fleshy grey, often large, caps; gills pale*

In rings in leaf-litter of deciduous and conifer woods. Late autumn. Common. Cap 5–15 cm (2–6 in), convex with inrolled margin, then expanding and broadly umbonate, finally slightly depressed; smooth, not fibrillose, often with a white, woolly "bloom"; pale grey to greyish fawn. Gills crowded, arcuate-decurrent; white then pale cream. Stem rather short, fibrous, soft; greyish with white swollen base. Spores cream, 5–8 × 3–5 μm. Usually considered edible but is suspected in some cases of mild poisoning.

2 *Clitocybe* (=*Lepista*) *flaccida*: *funnel-shaped caps, tan to rich orange; texture rather floppy, leathery*

In mixed woodlands in leaf-litter. Autumn. Frequent to common. Cap 5–8 cm (2–3 in), convex then rapidly funnel-shaped; soft, leathery, reddish tan to rich orange (this latter colour form is often called *C. inversa* and treated as a separate species). Gills crowded, deeply decurrent; pale, whitish to tan. Stem rather slender above, slightly swollen below, smooth; tan to orange. Spores pink, minutely prickly-warty, 4–5 × 3–4 μm. Edible but not very good. Usually considered a true *Clitocybe* but recent studies place this with the Blewits genus, *Lepista*.

3 *Clitocybe clavipes* (Club Foot): *umbonate to depressed, grey-brown cap; gills a distinct pale yellow; stem swollen, flaccid*

In leaf-litter or needles of mixed woods. Autumn. Common. Cap 4–10 cm (1½–4 in), convex then expanded and umbonate, slightly depressed with age; smooth, greyish brown to buff. Gills thick, soft, deeply decurrent. Stem very bulbous-clavate, soft, hairy at base, fibrillose; pale grey. Spores white, 4–5 × 3–4 μm. Edible but not recommended.

1 *Clitocybe geotropa:* large depressed but umbonate cap; tall stem

In woodland clearings and margins. Autumn. Occasional. Cap 10–20 cm (4–8 in), beginning convex with a small umbo then rapidly expanding to slightly funnel-shape still with umbo; pale buff to ochre, smooth. Gills rather crowded, decurrent; white then pinkish-buff. Stem longer than cap diameter, stout, fibrous, clavate; colour as cap, base downy-hairy. Spores white, 6–8 × 5–6 µm. Edible and delicious.

2 *Clitocybe infundibuliformis* (Common Funnel-cap): *thin, pale, funnel-shaped cap; slender stem*

In leaf-litter or grass in deciduous woodlands and heaths. Autumn. Frequent to common. Cap 3–8 cm (1–3 in), rapidly funnel-shaped; smooth, pale pinkish buff to yellow-ochre. Gills deeply decurrent, crowded; white to pale buff. Stem rather long, slender, swollen below, rather tough; pale buff. Spores white, tear-drop shaped, 6–7 × 3–4 µm. Edible and quite good. This rather attractive species is easily recognized by the thin, pale caps, pleasant slightly acidic smell and rather slender stature.

3 *Clitocybe odora* (Aniseed Toadstool, Anise Cap): *whole toadstool blue-green or pea-green and with clear strong smell of aniseed*

In leaf-litter of deciduous woods, especially on roadsides and paths. Autumn. Frequent. Cap 3–7 cm (1–3 in), convex then expanded and slightly umbonate, margin wavy, irregular; cuticle smooth, dry; green to blue-green. Gills narrow, adnate-decurrent, pale blue-green. Stem rather short, smooth, fibrillose; swollen and woolly at base; pale blue-green. Spores white, elliptic, 6–8 × 3–4 µm. Edible and delicious as flavouring or with small pieces added to a dish.

4 *Clitocybe rivulosa:* small white cap with delicate bloom usually in zones; stem fibrous, often twisted

In fields, roadsides and gardens in grass. Summer and autumn. Frequent to common. Cap 2–6 cm (1–2½ in), convex then expanded and slightly depressed, margin incurved; white to pale pinkish-tan. Gills crowded, thin, decurrent or adnate, white to cream. Stem usually shorter than cap diameter, slender, smooth; woolly at base; fibrous and often noticeably twisted; white to pale buff. No distinctive odour. Spores white, elliptic, 3–5 × 2–4 µm. *Dangerously poisonous, even deadly.*

The very similar *C. dealbata* found in similar localities is also *dangerous*; it has a lead-white, non-zoned cap, gills slightly yellowish, stem short, smell floury.

CLITOCYBE (contd)

1　*Clitocybe langei: small hygrophanous (changing colour when wet) grey-brown to pale tan cap; greyish gills; floury smell and taste*

In conifer woods and under bracken by birches. Late autumn to winter. Common. Cap 2–4 cm (1–1½ in), convex then expanded and slightly depressed; smooth, often bicoloured when partially dry; striate at margin when wet. Gills rather close, narrow, decurrent, greyish. Stem slender, smooth, pale, greyish tan. Spores white, tear-drop shaped, 5–7 × 2–3 μm. Not recommended for eating, as it is difficult to identify accurately.

The very similar *C. vibecina* differs only in spore morphology. *C. fragrans*, also pale tan-grey, has an odour of aniseed.

LEUCOPAXILLUS

2　*Leucopaxillus giganteus* (=Clitocybe gigantea): *very large white cap with downy margins; gills and stem also white; often in rings*

In often large rings in pastures, hedgerows, killing nearby grass. Autumn. Occasional to frequent. Cap 10–30 cm (4–12 in), soon funnel-shaped; margin inrolled, woolly-pubescent; ivory-white, often rather rough and cracked or scaly at very centre. Gills crowded, decurrent, forking. Stem short, stout, tough and fibrillose. Spores white, amyloid, 6–8 × 3–6 μm. Edible and recommended. This species is now removed from *Clitocybe* into the separate genus *Leucopaxillus* because of its differing anatomy and amyloid spores.

HYGROPHOROPSIS

3　*Hygrophoropsis aurantiaca* (False Chanterelle): *in heathy woods of birch and pine; orange-yellow cap and paler gills*

Summer and autumn. Abundant. Cap 3–6 cm (1–2½ in), convex then expanded and funnel-shaped, margin inrolled; slightly downy; bright orange-yellow to almost cream-colour when on open heaths. Gills crowded, thin, repeatedly forked, decurrent; orange or pale yellow; soft and easily separable from cap flesh. Stem soft, sometimes swollen, smooth; orange above and darker brown below. Spores white, elliptic, 7–8 × 4 μm, dextrinoid. Usually considered poisonous but actually edible although not very good. It is often mistaken for the true Chanterelle (*Cantharellus cibarius*, pp. 206–7), which however lacks true gills, having rounded "gill-like" wrinkles on the cap undersurface.

LACCARIA

1 *Laccaria proxima:* like the more common Deceiver but taller, stouter, more robust in all parts; spores elliptic

In rather damp, often boggy places. Autumn. Frequent. This variable species is similar in many respects to *L. laccata* but distinguishable by its different build, spores (7–10 × 6–8 µm), and habitat. It grows to 12 cm (4½ in) high. Edible and tasty.

2 *Laccaria laccata* (Deceiver): can be extremely deceptive, but reddish brown colours, fibrous stem and pinkish gills are typical

In troops in very mixed habitats from dark woodlands to paths or open heaths. Summer and autumn. Abundant. Cap 2–4 cm (1–1½ in), convex then expanded and slightly depressed; reddish brown to tan on drying; smooth to slightly scaly at centre. Gills thick, distant, adnate; pinkish to reddish brown, dusted with white spores when mature. Stem slender, tough, fibrous; colour as cap. Spores white, globose, spiny, 7–8 µm. Edible and quite tasty although rather small.

The very similar species *L. bicolor* has a reddish brown cap with lilac gills and bright lilac-violet woolly stem base; uncommon.

3 *Laccaria amethystea* (Violet Deceiver): unmistakable deep amethyst cap and stem (but paler when dry)

In damp, rather shady woods, on stream sides etc. Autumn. Common. Cap 2–3 cm (1 in), convex then expanded and slightly depressed; intense deep amethyst or violet when moist, but hygrophanous, so paler greyish-pink to bluish when dry. Gills deep violet, adnate, dusted with white spores when mature. Stem slender, fibrous, colour as cap. Spores white, globose, with tiny non-amyloid spines, 9–11 µm. Edible but rather too small for the table.

OMPHALINA

4 *Omphalina ericetorum:* cap small, top-shaped, with scalloped edge; stem apex darker; many species rather similar

In troops, on heathy soils, peaty areas in woods etc. Autumn. Common. Cap 0.5–2 cm (¼–1 in), convex then soon flattened and slightly depressed; edge scalloped, with radial grooves to centre; pale yellow-brown to olive. Gills white to yellowish, adnate-decurrent, distant. Stem slender, smooth, colour as cap; darker at apex, woolly at base. Spores white, elliptic, 8–10 × 5–6 µm. Edibility not known, and far too small to be of any importance.

COLLYBIA

1 *Collybia butyracea* (Greasy Cap, Butter Cap): *usually umbonate cap with "greasy" texture and darker umbo; stem club-shaped*

In mixed woods in leaf-litter. Autumn. Common. Cap 3–8 cm (1–3 in), convex then expanding with low umbo, smooth and very greasy when moist; greyish brown to reddish or olive fading when dry but usually with umbo and margin darker. Gills whitish, crowded, free from stem. Stem often rather swollen at base, tapering upwards, smooth but with woolly-hairy base; colour as cap. Spores white, elliptic, 6–7 × 3–4 μm. The odour is slightly rancid. Edible but not particularly delicious.

2 *Collybia dryophila*: *pale yellowish colours; thin stem, tough and flexible, with slightly woolly base*

In clumps in mixed woodlands or heaths. Autumn. Often abundant. Cap 2–4 cm (1–1½ in), convex then soon flattened, thin-fleshed, often wrinkled; colour very variable, reddish tan or pale buffy-tan to yellowish or even white. Gills white to distinctly yellowish; narrow, crowded, free to adnexed. Stem thin, tough but flexible, smooth and only slightly woolly-hairy at base; colour as cap but paler. Spores white, elliptic, 5–6 × 3–4 μm. Edible but small.

3 *Collybia distorta*: *cap brownish, rather large and broadly bell-shaped; stem twisted and grooved*

In groups in mixed woods, usually pine. Autumn. Occasional. Cap 5–9 cm (2–3½ in), convex then broadly umbonate, red-brown, rather thin-fleshed and soft; margin rather wavy. Gills very crowded, adnate, whitish with reddish brown spots or stains; margin uneven to slightly toothed. Stem smooth, cartilaginous, pale tan, often twisted and distorted, grooved and furrowed. Spores white, elliptic 5–6 × 4–5 μm. Edible but worthless.

4 *Collybia erythropus*: *strong contrast between very pale cap and deep red to brownish red stem; rather small slender species*

In deciduous woodlands in leaf-litter. Autumn. Frequent. Cap 2–3 cm (1 in), convex then soon flattened, smooth, dry, slightly wrinkled; pale buff-tan to almost white when dry. Gills free, not crowded, whitish to flesh colour. Stem slender, often flattened, smooth and shining. Spores white, pip-shaped, 6 × 2–3 μm. Edible but too small and thin to be worth collecting.

COLLYBIA (contd)

1 Collybia maculata (Red-spot Cap): *tough, sinewy, pure white with rust-brown spots and stains*

Often in large rings in leaf-litter of deciduous woods. Summer and autumn. Very common. Cap 4–10 cm (1½–4 in), convex then soon flattening; dry and smooth; pure white at first then soon spotted and stained with rust-coloured spots, finally completely reddish brown. Gills very crowded, free; edge minutely toothed; whitish then soon spotted reddish brown. Stem rather tall, slender to stout, smooth to fibrillose, very tough and often rooting; white then spotted like cap. Spores pinkish-cream, almost spherical, 4–5 × 5 µm. Not edible; very tough and rather bitter.

2 Collybia fusipes (Spindle-stem): *always clustered at base of deciduous trees or stumps, usually oak; swollen, spindle-shaped stem; red-brown colours*

Summer and autumn. Common. Cap 3–6 cm (1–2½ in), convex then slightly expanding with broad umbo; smooth, dry to slightly greasy in wet weather; dull brick-red to deep brown, paler tan when dry. Gills broad, thick, not crowded, edge often crinkled, with interconnecting "veins" between gills; whitish then pale brownish with darker reddish spots. Stem usually rooting, several fused together at base; very variable in length and width, often grossly swollen and spindle-shaped, splitting into deep cracks in dry weather; colour as cap only paler. Flesh very tough, pliant. Spores white, elliptic, 5–6 × 3–4 µm. To be avoided, as it is so tough it can remain *apparently* unchanged for many days, even weeks, but is then old, inedible and could possibly cause food-poisoning.

3 Collybia peronata (Wood Woolly-foot): *rather dull, tan cap; yellowish stem with thick woolly-hairy base*

In clumps in deciduous woods. Autumn. Common. Cap 3–6 cm (1–2½ in), convex then expanded and flattened, yellowish tan to earth-brown. Gills rather distant, pale yellowish, distinctly woolly-hairy at base often binding leaf-litter together. Spores white, pip-shaped, 7–9 × 3–4 µm. Has a distinctly peppery taste when chewed for a minute or two and is not considered to be an edible species.

The related *C. fuscopurpurea*, in beech woods and much less common, is distinguished by the dark reddish brown to deep purplish brown cap, and stem with yellower hairs on lower half.

COLLYBIA (contd)

1 Collybia confluens: *tall, slender, fused stems, in tufts; colours pinkish grey*

In dense tufts in mixed woods, often in large circles in leaf litter. Autumn. Common. Cap 2–4 cm (1–1½ in), convex then flattening; dry, very pale buff, flesh colour or almost white. Gills narrow, crowded, colour as cap. Stem tall, slender, rather tough and pliant, greyish-flesh to reddish, minutely velvety-pubescent; usually many stems fused together for about one-third of their length. Spores white, elliptic, 6–9 × 3–4 µm. Edible but not recommended.

MARASMIUS

2 Marasmius rotula: *black stem; but whitish, strongly grooved; gills attached to "collar"*

On decaying sticks and roots in mixed woodlands. Autumn. Common. Cap 0.5–1.5 cm (½ in), flattened with centre flattened and strongly radially grooved, whitish. Gills distant, connecting to "collar" around the stem apex like spokes of a wheel. Stem very slender, smooth, shining, black. Spores white, pip-shaped, 7–10 × 3–5 µm. Too fragile to consider for the table.

Two rather shorter, but equally small-capped species are *Micromphale foetidus*, stocky, dull grey-brown, on branches of beech, with unpleasant smell, and *Marasmius ramealis*, a tiny pinkish white species found in large numbers on twigs.

3 Marasmius androsaceus (Horse-hair Fungus): *long slender stems like horse-hair; reddish brown caps*

On twigs, needles and debris of conifers, also heather. Autumn. Frequent. Cap 0.5–1 cm (½ in), flattened with central depression, rather wrinkled with radial grooves; dull reddish brown. Gills rather narrow, adnate, sparse, colour as cap. Stem extremely slender, hair-like, black, shining. Spores white, pip-shaped, 6–9 × 3–4 µm. Far too fragile and small to consider worth eating. The stems arise directly from the thin, horse-hair-like, black mycelium which travels through and around the material on which it grows.

4 Marasmius oreades (Fairy Ring Toadstool): *buff-coloured, often umbonate caps; in rings in grass*

In large rings in lawns, fields, and grassy clearings or paths in woods. Summer and autumn. Common. Cap 2–5 cm (1–2 in), convex then soon expanded with a broad umbo, margin often slightly grooved particularly in old, wet specimens; ochre-tan to pinkish buff. Stem short to medium length, smooth, colour as cap but white, woolly, below. Spores white, pip-shaped, 9–11 × 5–6 µm. Edible and delicious, but care must be taken to avoid picking some of the poisonous *Clitocybe* species which can grow along with it such as *C. rivulosa* (pp. 132–3).

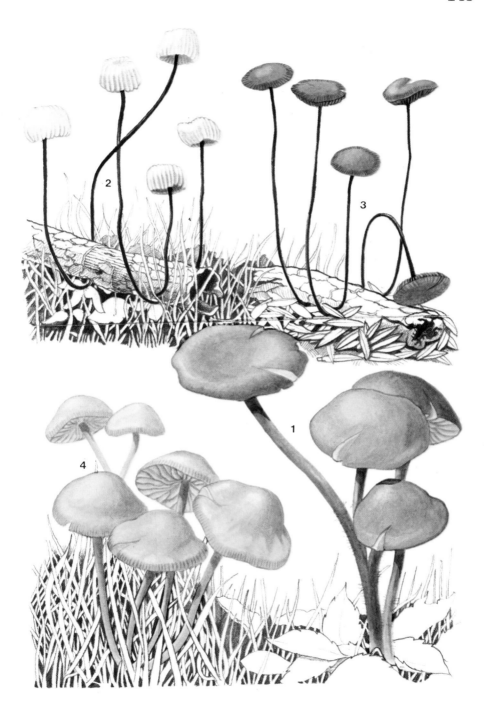

MARASMIUS (contd)

1 *Marasmius alliaceus* (Garlic Mushroom): *tall, black, slender stem; strong smell of garlic when bruised*

On leaf-litter and twigs of deciduous woods, especially beech. Autumn. Occasional. Cap 2–4 cm (1–1½ in), convex then soon expanding with slightly radially grooved margin; pale greyish to clay-brown. Gills whitish, adnexed, rather distant. Stem tall, slender, rooting; smooth and polished, black. Spores white, ovate, 9–12 × 6–7 μm. Has been used by some for garlic flavouring but not really recommended.

Similar is *M. scorodonius*, also with a garlic odour but smaller and with reddish brown colours, in grass.

FLAMMULINA

2 *Flammulina* (=Collybia) ***velutipes*** (Velvet stem, Winter Fungus): *appears in the winter months; cap yellowish; stem velvety, dark*

On dead or decaying deciduous timber. Common. Caps 2–6 cm (1–2½ in), convex then soon expanding, fleshy; bright yellowish to orange, darker at centre when wet, and with rather moist, slippery texture. Gills white to pale yellowish, adnexed, broad. Stems varying in length and thickness, densely tufted, curving upwards; rich reddish brown to almost black, and velvety at base. Spores white, elliptic, 7–10 × 3–4 μm. Edible and delicious; recommended. Appears when few other fungi are present.

TRICHOLOMOPSIS

3 *Tricholomopsis* (=Tricholoma) ***rutilans:*** *vividly coloured yellow cap masked by purplish squamules, and gills yellow*

On stumps and logs of conifers. Summer and autumn. Common. Cap 4–12 cm (1½–4½ in), convex then expanded, rich golden-yellow overlaid with dark purplish flecks and scales, darker and more densely scaled at centre. Gills broad, distant, adnexed, golden-yellow. Stem coloured as cap undersurface but paler and less squamulose; rather stout, without mycelial strands. Spores white, ovate, 5–7 × 4–5 μm. Apparently edible but not recommended. This beautiful and unmistakable species is similar to *T. decora*, which is also found on conifers, but in northern regions, and is rather paler with brownish, not purplish, squamules.

4 *Tricholomopsis* (=Collybia) ***platyphylla:*** *cap and stem greyish, fibrillose; gills distant; mycelial strands at base prominent*

On stumps, logs or buried timber of deciduous wood, attached by bootlace-like white mycelial strands. Summer and autumn. Common. Cap 4–10 cm (1½–4 in), soon flattened; smooth, dry, radially fibrillose, dull greyish to yellowish brown. Gills broad, very distant, adnexed to free, whitish. Stem rather short, straight and thick; dry, fibrillose, colour as cap or whitish. Spores white, 6–8 × 6–7 μm. Not edible because of toughness and rather bitter flesh.

ARMILLARIA

1 ***Armillaria*** *(=Clitocybe)* **tabescens:** *similar to the Honey Fungus but stems not bulbous, no ring, denser tufts*

On deciduous timber especially oaks. Late summer to early autumn (almost always *before* the true Honey Fungus). Uncommon except in warmer years, more frequent in southern parts. Remarkably similar to *Armillaria mellea* but more densely tufted, stems fused, not bulbous, entirely without ring. Spores white, ovate, 8–10 × 5–7 μm. Edible when cooked.

2 ***Armillaria mellea*** (Honey Fungus, Honey Tuft): *tufted on wood; bulbous stems with thick ring; scaly cap centre*

On a wide variety of coniferous and deciduous timber and shrubs, usually in large clumps; a serious parasite. Autumn. Abundant. Cap 4–12 cm (1½–4½ in), convex then expanded with low umbo, minutely scaly, especially at centre; colour variable, pinkish tan to ochre-brown or reddish, scales darker brown. Gills whitish, then pinkish brown often spotted darker brown; adnate or slightly decurrent. Stem short, bulbous, soon lengthening; colour as cap or paler, whitish. Ring conspicuous, thick, yellow, with scales at margin and below. Black bootlace-like mycelial strands attached to stem bases travel throughout stump, under tree bark etc. Spores pale cream, elliptic, 8–9 × 5–6 μm. Flesh smells rather strong, taste often very bitter, burning. Edible when cooked (removes bitterness), considered delicious by many. Now thought to consist of several species distinguished mainly by colour, scale development and spore-print. Keep careful notes on any different forms you may find.

OUDEMANSIELLA

3 ***Oudemansiella mucida*** (Porcelain Fungus): *unmistakable glistening white caps on beechwood; short stem with thin ring*

Often in large numbers on beech trunks and logs. Autumn. Frequent. Cap 3–6 cm (1–2½ in), convex then flattened with umbo; pure, glistening white to greyish, extremely slimy. Gills white, broad, distant, adnexed. Stem slender, tough, whitish to grey, fibrillose. Spores white, ovate, 13–18 × 12–16 μm. Edible but not recommended.

4 ***Oudemansiella*** *(=Collybia)* **radicata** (Rooting Shank): *very tall straight stem, deeply rooting; slimy wrinkled cap*

Around stumps or buried timber. Autumn. Common. Cap 3–8 cm (1–3 in), convex to bell-shaped then expanding with central umbo, margin wrinkled; very slimy when wet; clay-brown to ochre, darker at centre, radially wrinkled. Gills white, broad, distant, adnexed, edge often brown. Stem soon very tall, to 20 cm (8 in), slender, tapering upwards; whitish, darker below, fibrillose, striate, and with strong "taproot". Spores white, ovate, 12–6 × 10–12 μm. Edible but not recommended. The similar but rare *O. longipes* is darker brown with velvety, not slimy, cap and stem.

MYCENA

1 Mycena acicula (Orange Bonnet): *tiny bright orange cap with yellow stem*

On leaf-litter and twigs in deciduous woods. Autumn. Frequent. Cap 0.25–0.5 ($\frac{1}{8}$–$\frac{1}{4}$ in), convex, orange-red to scarlet, minutely striate. Gills orange-yellow, edge whitish; broad, distant, adnate to free. Stem long, slender, smooth, bright yellow, fading, base hairy. Spores oblong-fusiform, white, 9–12 × 2–4 µm. Edibility: like all *Mycena* species, too small to consider as food. Two other species may be confused with *M. acicula*: *M. coccinea*, on conifer twigs, generally redder in all parts, especially stem, and *M. adonis*, rather rare, cap pinkish, stem white, in grassy clearings in woods.

2 Mycena alcalina: *dark greyish colours; strong smell of domestic bleach*

Clustered on stumps or buried wood of coniferous trees. Autumn. Common. Cap 1–4 cm ($\frac{1}{2}$–$1\frac{1}{2}$ in), convex-conical then expanding, dark grey-brown to almost black; smooth, deeply striate when moist. Gills whitish to grey, often with darker margin, adnate. Stem slender, smooth, shining, yellowish grey; hairy at base and usually rooting, in clumps. Spores white, elliptic, 8–10 × 6–7 µm.

3 Mycena epipterygia: *cap and stem slimy, yellowish green*

On litter of conifer woods or heaths. Summer and autumn. Common. Cap 1–2 cm ($\frac{3}{4}$ in), convex then bell-shaped; smooth, slimy with striate margin: pale yellowish brown. Gills white, adnate. Stem slender, smooth and slimy; bright yellow to greenish yellow. The slimy skin is easily removed. Spores white, elliptic, 8–11 × 4–5 µm.

4 Mycena galericulata (Bonnet Mycena): *cap pinkish grey, relatively large; gills pinkish with cross-veins*

In clumps on deciduous wood. All year. Abundant. Cap 2–5 cm (1–2 in), conical to bell-shaped then expanded with umbo; very variable in colour from pale grey to pinkish brown or olivaceous; striate at margin. Gills white to pinkish, rather thick with noticeable cross-veining at base. Stem long, slender, smooth; hairy at base, rooting; colour as cap or yellower. Spores white, elliptic, 10–11 × 6–8 µm.

5 Mycena crocata: *dull grey-brown cap with saffron stem containing red latex*

In leaf-litter and twigs in beech woods. Autumn. Occasional. Cap 1–2 cm ($\frac{3}{4}$ in), conical then expanded, campanulate; greyish brown to olive, often paler, almost white; smooth with slightly striate margin. Gills white, adnate. Stem long, slender, smooth and shining; base rooting and hairy; paler above. Spores white, elliptic, 9–11 × 6–7 µm.

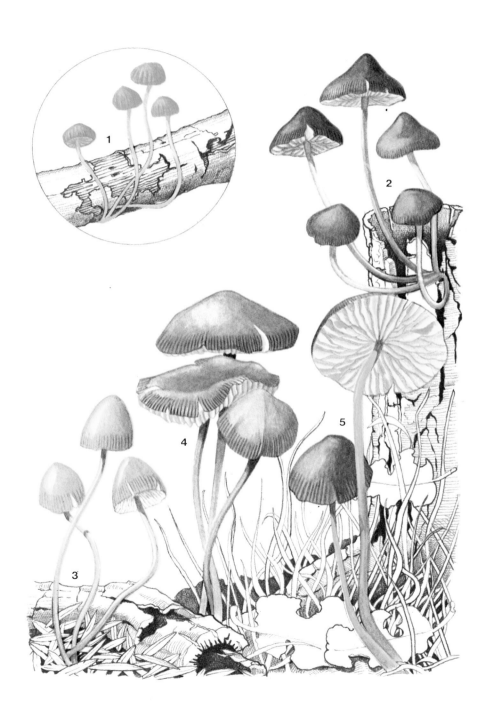

MYCENA (contd)

1 ***Mycena inclinata:*** *always on oaks; brown cap with dentate margins; stem reddish below, white above; strong odour when fresh*

In large clumps on old or dead oak trees and stumps. Autumn. Common. Cap 1–3 cm ($\frac{1}{2}$–1 in), convex then expanded; strongly striate; margin dentate, protruding beyond gills; pale to rich brown, becoming paler, almost white at margin. Gills whitish to grey, adnate, crowded. Stem long, slender, smooth; rich reddish brown and hairy at base fading to white in upper half. Odour often strong, rancid or soapy. Spores white, subglobose, 8–10 × 6–8 μm. Edibility: like all *Mycena* species, too small to be of culinary interest.

2 ***Mycena polygramma:*** *steel-grey colours; stem with faint white longitudinal raised lines*

In clusters or solitary on deciduous stumps and logs, buried wood. Autumn. Frequent. Cap 2–4 cm (1–1$\frac{1}{2}$ in), bell-shaped then expanded with umbo, margin striate; dark steel-grey, slightly yellowish with age. Gills white to greyish often stained pink. Stem long, slender, rooting, smooth and rigid; steel-grey with minute, white, raised longitudinal lines. Spores white, elliptic, 9–12 × 6–8 μm.

3 ***Mycena haematopus*** (Bleeding Mycena): *stem yields reddish latex; gills whitish, not red-edged*

In clumps on stumps of deciduous trees. Autumn. Frequent. Cap 1–3 cm ($\frac{1}{2}$–1 in), bell-shaped, greyish brown sometimes with white powdery bloom, striate at margin. Gills white then flesh-tinted to deeper reddish brown (red-edged in the similar *M. sanguinolenta*, p. 151), adnate. Stem slender, rigid, fragile; colour as cap darkening below to reddish brown; when broken yields red-brown latex. Spores white, elliptic, 10 × 6 μm.

4 ***Mycena galopus*** (Milk-drop Mycena): *grey to pure white colours; stem yields white latex*

In leaf-litter, pine needles, twigs etc., of mixed woods. Summer and autumn. Common. Cap 1–2 cm ($\frac{1}{2}$–1 in), convex to bell-shaped then slightly expanding; strongly striate almost to centre; from dark greyish brown to almost white in the variety *alba*; centre darker. Gills whitish, adnexed. Stem slender, smooth, polished, slightly hairy at base, colour as cap; when broken exudes drops of white, milk-like latex. Spores white, oblong-elliptic, 12–14 × 6–7 μm.

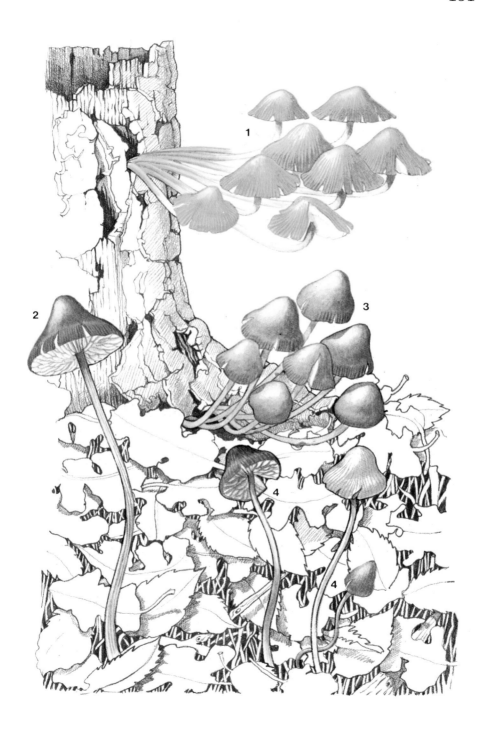

MYCENA (contd)

1 ***Mycena pura*** (Lilac Mycena): *lilac-pink colours and strong smell of radish*

In leaf-litter in mixed woods. Autumn. Common. Cap 2–4 cm (1–1½ in), conical then expanded with umbo; smooth, margin striate; colour very variable, white, violaceous, lilac to pink (as shown) fading to greyish with age. Gills white to pinkish, adnate. Stem rather stout, hollow, coloured as cap; base woolly. Flesh with strong odour of radish when bruised. Spores white, ovate, 6–8 × 3–5 µm. Like other *Mycena* species, not of culinary interest.

The similar beech-wood species *M. pelianthina* also smells of radish and is of similar appearance but has gill edges coloured dark purple.

2 ***Mycena sanguinolenta:*** *small, with reddish brown colours; red latex in flesh; gill edge red*

Usually in troops, not clumps, among mosses and fallen leaves. Autumn. Common. Cap 0.5–1 cm (¼–½ in), convex to bell-shaped, striate at margin; reddish brown, darker at centre. Gills whitish or coloured as cap but paler; edge dark-red when viewed under hand-lens (compare with *M. haematopus*, p. 150). Stem very slender, colour as cap but paler; yields thin reddish latex when broken. Spores white, elliptic, 8–9 × 4–6 µm.

3 ***Mycena vitilis:*** *very small, tall and slender; cap sharply umbonate*

In leaf-litter in deciduous woods, usually not tufted. Autumn. Common. Cap 0.5–1.5 cm (½ in), conical then expanded with prominent umbo, striate to centre; grey-brown. Gills greyish, thin, adnate. Stem long, very slender, smooth; very pale, almost white; base rooting, binding leaf-litter. Spores white, elliptic, 9–12 × 5–7 µm.

MACROCYSTIDIA

4 ***Macrocystidia cucumis:*** *pinkish cinnamon to red brown; flesh with strong fishy or cucumber odour*

On bare soil in mixed woods, especially on pathsides. Autumn. Occasional. Cap 2–5 cm (1–2 in), conical-campanulate, smooth, margin slightly striate; often pruinose. Gills crowded, emarginate, yellowish flesh colour. Stem tough, velvety, short to medium length; deep reddish brown to almost black, paler at apex. Spores pale reddish brown, oblong-elliptic, 8–10 × 3–4 µm.

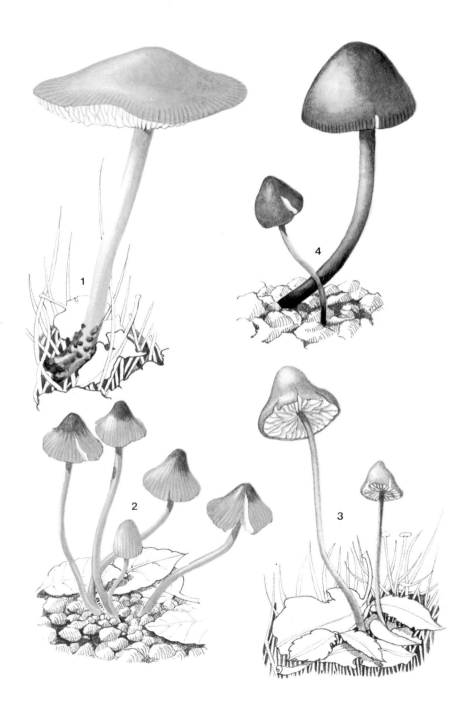

PLEUROTUS

PLEUROTACEAE

This family consists almost entirely of fungi growing on wood, and having usually rather tough, fleshy fruit-bodies with either no stem or a rather off-centre (eccentric) stem. The spores are white to pale cream or pinkish. Some species are well known edibles and one in particular, commonly called Shiitake (*Lentinodium edodes*) is widely cultivated throughout the Orient.

1 ***Pleurotus dryinus*** *(=corticatus): cap convex, felty-woolly with scales; stem with ring zone*

On elms, usually old living trees rather than stumps. Autumn. Frequent. Cap 5–10 cm (2–4 in), convex then expanding, margin inrolled; flesh thick, surface soft and downy, separating into woolly flattened scales; pale greyish brown to cream. Gills deeply decurrent down to ring zone; thin, crowded, white then yellowish. Stem short, stout, usually off-centre (eccentric), with distinct veil at apex covering gills when young, soon rupturing to form ring zone; colour whitish. Spores white, oblong-cylindrical, 12–14 × 4–5 µm. Apparently edible but not recommended here.

2 ***Pleurotus ulmarius:*** *pale cap; stem long, rather off-centre; usually on elms*

On deciduous trunks. Autumn. Occasional. Cap 5–20 cm (2–8 in), convex then expanding, thick-fleshed; smooth but often cracking into irregular scales; white then pale ochre. Gills sinuate, whitish to pale ochre. Stem variable in length, often quite long, to 5–15 cm (2–6 in); thick, rather eccentric or almost laterally attached to the cap; whitish, base often tomentose. Spores white, globose, 5–6 µm. Not recommended; best avoided. This species is now regarded by many as a Lyophyllum (*L. ulmarium*).

3 ***Pleurotus ostreatus*** (Oyster Mushroom): *shell-shaped fruit-bodies; cap smooth, bluish when young*

In large numbers on various deciduous trees (especially beech). Summer and autumn. Abundant. Cap 5–15 cm (2–6 in), button-shaped when very young then soon expanding and flattening to form flat, oyster-shell-shaped brackets, deep brown to bluish black when young but soon very pale buff to almost white with age. Gills deeply decurrent, thin, crowded, whitish cream. Stem absent or very short (except in var. *salignus*) white, woolly. Spores pale lilac in thick deposit, elongated, 10–11 × 3–4 µm. Edible and highly favoured by many.

The variety *salignis* appears late in the year and has a long hairy stem, deep bluish grey cap usually not fading, and gills also bluish. Considered a good species by some.

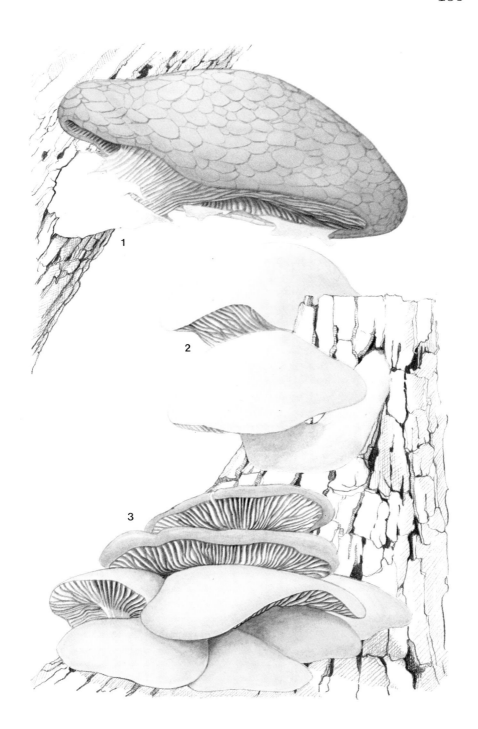

PANELLUS

1 *Panellus serotinus: olive-green to brownish with age, felty-tomentose; gills yellow; stem lateral*

On trunks, fallen logs, etc. of deciduous trees, especially beech. Autumn. Uncommon. Cap 3–10 (1–4 in), convex with inrolled margin then expanding, flattened, kidney-shaped to almost circular with lateral short stem; surface tomentose to smooth with age or when wet; deep green to yellow-green, then brownish. Gills adnate, crowded, narrow, yellow. Stem short, thick, tomentose; yellowish with brown squamules, darker near cap. Spores white, cylindric-elongate, 4–6 × 1–2 μm, amyloid. Edibility doubtful; to be avoided.

2 *Panellus (=Panus) stipticus: flattened, kidney-shaped, pale-clay caps; gills ochre; taste bitter*

On dead wood of deciduous trees. All year. Common. Cap 1–5 cm ($\frac{1}{2}$–2 in), convex then flattened, kidney-shaped to semicircular with lateral stem. Surface at first slightly pruinose-mealy then smooth; pale buff to clay-brown. Gills crowded, adnate, thin; pale ochre to cinnamon. Stem short, thin, tapering downwards, whitish. Spores white, elliptic, 4–5 × 2–3 μm, amyloid. Not edible.

PANUS

3 *Panus torulosus: very tough flesh; downy cap and stem with lilac flush when young*

On stumps of deciduous trees. Autumn. Common. Cap 5–10 cm (2–4 in), often very irregular, lobed, wavy at margin, convex then soon funnel-shaped; very tomentose, downy; pale ochre with clear lilac-violet tints on cap and stem when young, soon entirely ochre. Gills decurrent, narrow, crowded; pale ochre-brown often with lilac flush, soon flesh-colour. Stem very short to almost absent, often lateral; tomentose, colour as cap. Flesh tough, woody, often drying extremely hard. Spores white, cylindrical, 5–6 × 3–4 μm. Not edible.

LENTINUS

4 *Lentinus lepideus: tough, broadly scaly cap; pale colours; edges of gills serrate*

On old coniferous timber, occasionally in houses. Autumn. Occasional. Cap 5–12 cm (2–5 in), convex then slightly depressed at centre; skin cracking into broad, irregular scales; pale ochre or whitish yellow to brownish with scales darker, brown. Gills decurrent to slightly sinuate, distant, broad; with edge ragged and serrate; pale whitish yellow. Stem short to medium but rather variable in length and thickness; tough and fibrillose, often scaly-squamulose; colour as cap. Spores white, elliptic, 10–15 × 4–6 μm. Not edible.

The closely related *L. tigrinus* has a more depressed, funnel-shaped cap, thinner flesh, densely but minutely squamulose-scaly cap and rather slender, tapering stem with ring-like zone at apex; found on deciduous wood, uncommon.

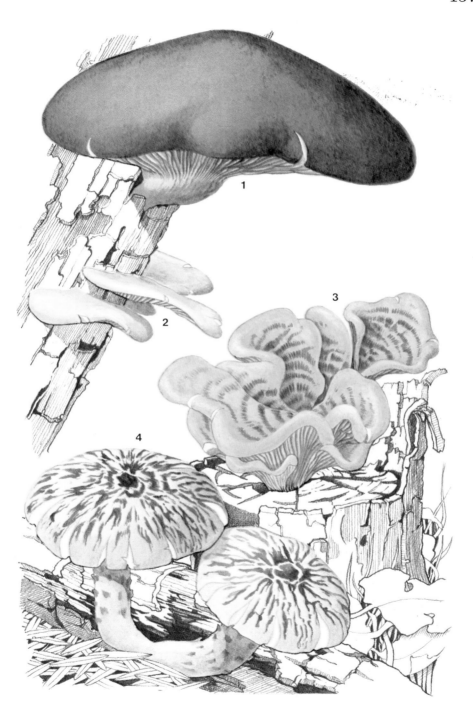

BOLBITIACEAE

This group consists, according to recent discussions, of three genera which have smooth medium-brown spores with a prominent germ-pore, and the cells of the cap cuticle rounded. Many species have a ring, most are small, even tiny. The robust *Agrocybe cylindracea* has for centuries been cultivated on poplar stumps and poles for food.

AGROCYBE

1 *Agrocybe erebia*: dull brown colours (including spore print); thin ring on stem

On bare soil, sometimes leaf-litter, in deciduous woods. Autumn. Frequent. Cap 3–6 cm (1–2½ in), convex then flattening with low umbo, margin often striate, deep umber when moist, soon drying paler, clay-brown. Gills adnate, whitish, soon cigar-brown. Stem rather stout, long, slightly fibrillose; whitish brown then darker. Ring prominent, pendent, grooved above, vanishing with age. Spores elliptic, 10–13 × 5–7 μm. Apparently edible but best avoided as easily misidentified. *A. praecox* is of similar build and stature; *A. dura*, in spring, is stouter, tougher, the cap often cracking; both are common.

2 *Agrocybe cylindracea* (=Pholiota aegerita): smooth but often wrinkled and cracked; ring on stem; in clumps on wood

On stumps and trunks of poplars, elms, occasionally other deciduous trees. Autumn. Occasional. Cap 4–10 cm (1½–4 in), convex then slightly expanded; silky surface often wrinkled and channelled, centre cracking; pale ochre-buff to clay-brown, paler, almost white at margin. Gills adnate-sinuate with decurrent tooth; pale ochre then dull cigar-brown. Stem fibrous, tapered downwards, white to pale clay. Ring high, thin, membranous, soon torn and vanishing. Spores dull brown, elliptic, 9–10 × 5–7 μm. Edible and delicious.

CONOCYBE

5 *Conocybe tenera*: tall slender stem and conical cap; yellow-orange colours; spores pale rust-brown

In grass in fields, woods and roadsides. Summer and autumn. Common. Cap 2–4 cm (1–1½ in), conical then slightly expanded-campanulate; smooth, ochre to orange-brown. Gills narrow, adnate (to free), crowded, cinnamon. Stem smooth, colour as cap. Spores oval, 10–12 × 5–7 μm. Not edible. The related *C. lactea*, uncommon, in grass, can be easily identified by its milk-white cap and stem and cinnamon gills.

BOLBITIUS

4 *Bolbitius vitellinus*: bright egg-yellow cap, thin, viscid

On horse dung, manured grass, sometimes wood chips. Summer and autumn. Common. Cap 2–4 (1–1½ in), deeply campanulate-conical, soon expanded, flattened; margin striate; thin-fleshed, rapidly collapsing. Gills free, thin, crowded, pale cinnamon. Stem tall, slender, white to pale yellow, with woolly-powdery coating. Spores rust-brown, elliptic, 11–15 × 6–9 μm. Too small for eating.

HYGROPHORACEAE

This group contains some of our most colourful and spectacular species. Bright reds, yellows, greens and orange are the rule rather than the exception in this family, which also includes pure white and the more usual greys and browns. Cap and stem are often very viscid. All species have thick, waxy gills, white spores and long basidia. The genus *Hygrophorus* was formerly divided into three groups: *Limacium*, with slimy cap, adnate-decurrent gills and slimy stem usually dotted at the apex; *Camarophyllus*, with dry cap, smooth, fibrous stem and gills free to decurrent; and *Hygrocybe* with thin, fragile, moist to viscid cap, smooth stem and gills free to decurrent. The last two are now considered to form the single genus *Hygrocybe*, usually associated with open grassland; *Limacium* now forms the genus *Hygrophorus*, associated with trees. Despite the often bright colours, the hygrophori can be surprisingly difficult to identify. Recent studies have shown the existence of many more species than formerly recognized; *precise* notes of colours, viscidity, taste etc. are necessary for accurate identification.

HYGROPHORUS

1　*Hygrophorus hypothejus* (Winter Hygrophore): *cap slimy, olive-brown; gills yellow; stem pale; late in season*

In pine woods. Late autumn to early winter. Common. Cap 2–6 cm (1–2½ in), convex then expanding, depressed; radially fibrillose. Gills decurrent, brown and rather sparse, thick, waxy, yellow then saffron. Stem cylindric, whitish or pale yellowish olive, slimy below ring-like zone. Spores white, ovate, 7–9 × 4–5 µm. Edible but not recommended.

2　*Hygrophorus chrysaspis*: *slimy white cap and stem discolouring yellowish to reddish brown; specific chemical test*

In beech woods. Autumn. Frequent. Cap 2–5 cm (1–2 in), convex then expanded with low umbo; smooth and viscid. Gills adnate-decurrent, thick, waxy; whitish often with a pale orange-pink hue, discolouring as cap. Stem fleshy, white, slimy; apex dotted with white flocci. Odour strong, fragrant to slightly sickly. Spores elliptic, white, 7–9 × 4–6 µm. Chemical test: flesh + potassium hydroxide = red-brown. Not known to be poisonous but not recommended. The very similar *H. eburneus* is found principally under oaks and lacks odour; flesh does not discolour; no reaction to potassium hydroxide.

3　*Hygrophorus chrysodon*: *slimy white cap and stem, with bright yellow flocci on cap edge and stem apex*

In mixed woods. Autumn. Uncommon. Cap 2–6 cm (1–2½ in), convex then expanded, broadly umbonate; smooth, very slimy; margin often overlapping gills, usually irregular. Gills adnate-decurrent, thick, waxy; white, margins often yellowish. Stem slender, often hollow; viscid. Odour strong, fragrant. Spores white, ovate, 7–9 × 4–6 µm. Edible but not recommended.

HYGROPHORUS (contd)

1 ***Hygrophorus bresadolae:*** *cap slimy, golden orange to reddish: among larches*

In grass near larches. Autumn. Uncommon. Cap 2–4 cm (1–1½ in), convex then expanded with low umbo; very viscid, bright golden orange to reddish orange. Gills adnate-decurrent, distant, whitish to pale yellow. Stem equal, white, pruinose above, pale tawny yellow below and viscid up to faint ring-like zone. Spores white, elliptic, 8–10 × 5–6 µm. Edibility: not recommended.

The very closely related and almost identical *H. speciosus*, frequent in northern parts of the U.S.A., differs mainly in its more intensely reddish orange cap and rather robust build.

HYGROCYBE

2 ***Hygrocybe*** *(Camarophyllus)* ***pratensis:*** *cap yellow-buff, top-shaped; gills deep, decurrent, with "cross-veins"*

In open fields, pastures. Autumn. Common. Cap 2–6 cm (1–2½ in), convex then expanded; dry, smooth, dull yellowish-buff to very pale buff. Gills decurrent, deeply so with age, giving the top-shaped appearance; colour as cap; thick and distant, often interconnected by "veins" at base. Stem rather stout, tapered at base, paler than cap. Spores white, elliptic, 7–8 × 5 µm. Edible and quite good.

3 ***Hygrocybe obrussea:*** *bright golden yellow, viscid cap and stem; obtusely conical; in woodland clearings*

In grass in woodlands. Spring and autumn. Uncommon. Cap 5–10 cm (2–4 in), obtusely conical to campanulate, slightly expanding with age; very viscid when moist; bright golden yellow, paler with age and margin often splitting. Gills white tinged pale yellow at base; soft, waxy. Stem medium to stout, often flattened and compressed; slightly tapered at base; slightly fibrillose, viscid; colour as cap or paler, white at base. Spores white, elliptic, 8–9 × 5–6 µm. Edible but not recommended.

4 ***Hygrocybe*** *(Camarophyllus)* ***nivea:*** *cap small, white, top-shaped; no odour; in fields*

In open fields, pastures, woodland clearings. Autumn. Frequent. Cap 2–5 cm (1–2 in), convex, broadly domed, then expanded, top shaped; dry, smooth, pure white to ivory; margin slightly striate. Gills decurrent, distant, white. Stem slender, tapered below, white. Spores white, elliptic, 7–8 × 4–5 µm. Edible and quite good.

The rather similar *H. russocoriacea* has a strong smell of leather, more ivory cap and larger spores.

HYGROCYBE (contd)

1 *Hygrocybe calyptraeformis* (Pink Wax-cap): *acute cap and pinkish colours*

In grassy meadows, pastures, woodland margins. Autumn. Occasional. Cap 2–5 cm (1–2 in), margin expanding, often splitting into 3–4 large wings; smooth, dry, slightly fibrillose; clear, pale rose-pink. Gills pale rose then whitish, waxy; distant, adnexed, very narrow at stem. Stem tall, graceful, fragile, striate-fibrillose, white to pale pinkish. Spores white, elliptic, 7–8 × 4–5 µm. Edible but not recommended.

2 *Hygrocybe chlorophana* (Golden Wax-cap): *very viscid cap and stem; constant yellow colours, especially in gills*

In fields and pastures. Autumn. Frequent. Cap 2–7 cm (1–3 in), obtusely conical; very viscid; bright lemon or chrome-yellow. Gills free to adnate, distant; waxy; whitish to yellow, deeper yellow near cap flesh. Stem fragile, very viscid (persistently); smooth, bright yellow. Spores white, elliptic, 6–9 × 4–6 µm. Edible but not recommended. Can be distinguished from several very similar yellowish hygrophori by the lemon-yellow gills, cap *and* stem without orange flush (sometimes a faint trace when young) and the persistent viscidity. *Hygrocybe obrussea* (pp. 162–3) has a soon dry, rather fibrillose cap, yellow with orange-brown tints, and dryish stem soon discolouring orange-brown.

3 *Hygrocybe coccinea* (Common Scarlet Wax-cap): *cap rather bell-shaped; deep blood-red to scarlet in all parts*

In fields, pastures, woodland margins. Autumn. Common. Cap 2–5 cm (1–2 in), moist then soon dry, smooth, convex to slightly obtuse; fading to orange or dull ochre. Gills adnate-decurrent, broad, waxy; with paler, orange margin. Stem rather slender, smooth, dry, slightly striate; paler, orange-ochre below. Spores white, elliptic, 7–9 × 4–5 µm. Edible and quite good. Similar species have more yellow or white in cap, stem or gills, or different cap shapes. *H. reai* has a convex to acutely conic cap, scarlet then orange, gills deep yellow flushed orange at the base, and taste bitter, not mild.

4 *Hygrocybe conica* (Blackening Wax-cap): *acute cap; yellow-orange colours; white to pale yellow gills; all parts blackening*

In pastures and fields, rarely in woods. Summer and autumn. Common. Cap 1–5 cm (½–2 in), acutely conical, then expanded but retaining acute umbo; dry, fibrillose; orange or yellowish, sometimes with red flush; strongly blackening all over. Gills free to adnexed, waxy; white to pale yellow before greyish, then black (never flushed red). Stem equal, slender, fibrillose; sometimes with red flush; blackening. Spores white, elliptic, 7–9 × 4–5 µm (3 µm larger on a frequent two-spored form). Edible but not highly recommended. *H. nigrescens* has a more obtuse

HYGROCYBE (contd)

cap and stem, both redder, the latter with a white base, often in woods as well as fields. In grass near the sea is found *H. conicoides*, very similar but with gills distinctly yellow, flushed red, also blackening.

1 Hygrocybe punicea (Large Crimson Wax-cap): *large obtuse, irregular cap; deep red colours, soon fading*

In woods, fields and hedgerows. Autumn. Occasional. Cap 5–10 cm (2–4 in), obtusely conical to bell-shaped, margin often lobed and irregular; smooth, slightly viscid; deep blood-red to cherry or crimson, soon fading to orange then pale yellow, finally whitish. Gills blood-red to purplish with yellow margin, then fading. Stem tall, rather stout, slightly viscid, fibrillose; red then soon yellowish but base persistently white. Flesh within stem also white. Spores white, elliptic, 8–12 × 4–6 μm. Edible and quite good. A large and striking fungus, not often found perfectly red, usually fading very rapidly to orange or yellow.

H. splendissima differs in the yellow flesh in its stem, stem base rarely white, and cap a more carmine red.

2 Hygrocybe psittacina (Parrot Toadstool, Parrot or Green Wax-cap): *very glutinous cap and stem; green coloration (persistent at stem apex)*

In short turf in fields and pasture. Autumn. Frequent. Cap 2–5 cm (1–2 in), convex then expanded with broad umbo; smooth, very viscid with thick gluten; at first bright green because of the gluten but as this disappears changing to yellowish or reddish, finally purplish. Gills adnate, broad, thick; green, then yellowish with green at base. Stem slender, shortish; persistently green at apex and all over at first, then yellowish below; viscid, tough. Spores white, elliptic, 7–10 × 4–6 μm. Edible but too slimy. A very distinctive and beautiful species of remarkable coloration.

3 Hygrocybe miniata: *flattened or depressed cap with tiny scales; in grass*

In heathland, pastures or fields. Autumn. Uncommon. Cap 1–2 cm ($\frac{1}{2}$–1 in), convex then flattening, often slightly depressed; dry, fibrillose or scurfy, soon torn into minute recurved scales especially at centre; deep blood-red to scarlet, then orange or ochre. Gills adnate, thick, orange-red to scarlet with yellow margin. Stem equal, slender, smooth; deep red, paler at base. Spores white, elliptic-oblong, 7–10 × 5–6 μm. Not of culinary interest.

The similar *H. turunda* grows in wetter, boggy places in sphagnum tussocks, it has more orange or yellow cap colours, darker brownish scales at the centre; margin fringed or dentate.

RUSSULACEAE

All Russulaceae have heteromerous flesh, the cells being a mixture of round and long cells, the latter usually containing latex. All have spores with strong ridges and networks on the surface which stain blue-black in iodine. The two principal genera, *Russula* and *Lactarius*, have, as a result of the mainly spherical cells which make up the flesh, a distinctive granular, crumbly texture. The flesh of all *Lactarius* species when broken exudes a coloured or milky latex. Many species are widely sought after for food while others can be extremely hot to the taste when raw. Tasting a small (pea-sized) piece of flesh or gill or a drop of "milk" is an essential part of identification. All russulas should be checked with iron sulphate ($FeSO_4$) as a matter of course. Very few species are poisonous, although several are very indigestible or emetic if eaten in quantity.

RUSSULA

1   Russula aeruginea (Grass-green Russula): *cap grass-green fading to yellowish; creamy yellow gills and spores*

In deciduous woods. Autumn. Common. Cap 5–10 cm (2–4 in), convex then flattened, dry. Gills yellow-buff, smooth, brittle, often forking. Stem equal, fairly solid, white. All parts brown-spotted with age. Spores broadly elliptic, 6–10 × 5–7 μm. Taste mild to slightly hot. Edible but not recommended. The similar *R. heterophylla* has white gills and spores.

2   Russula alutacea: *purplish cap and rose base of white stem; deep yellowish gills and spores; taste mild*

In deciduous woods. Autumn. Occasional. Cap 7–12 cm (3–5 in), convex then flattened, deep purplish or vinaceous-brown, centre often buff or olive. Gills distant, thick, strongly interveined. Stem rather stout, firm. Spores ovate, 8–10 × 6–9 μm. Chemical test: flesh + phenol = deep blackcurrant-juice colour (most *Russula* give a purple-brown). Edible and good. The closely related *R. olivacea* (same phenol reaction) has the stem *apex* rose, and cap more greenish-purple.

3   Russula atropurpurea: *cap deep purplish red with blackish centre; robust, firm build*

In mixed woods. Autumn. Abundant. Cap 4–10 cm (1½–4 in), convex then expanded, may be red or violaceous; often with very pale, cream blotches. Gills slightly decurrent, rather crowded, pale cream; frequent, fine rust-like spots on edge. Stem white tinted brownish, rust or grey when damp and old. Flesh firm then spongy. Spores off-white, ovate, 7–9 × 6–7 μm. Odour fruity of apples. Taste mild to slightly hot. One of the easiest species to identify. Edible and good.

4   Russula violeipes: *cap usually yellowish; stem purplish; spores cream*

In deciduous or conifer woods. Autumn. Occasional. Cap 4–8 cm (1½–3 in), convex then slightly expanded, and finally

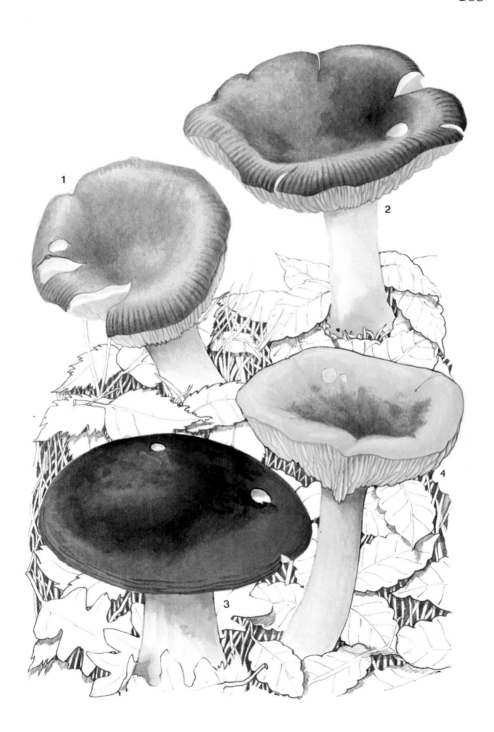

RUSSULA (contd)

depressed; bright lemon-yellow to straw, sometimes with reddish or purplish areas. Gills narrow, crowded, rather greasy; yellowish cream colour. Stem equal, white, usually with flush of yellow and purplish violet. Odour very distinct, fruity to unpleasant, shrimpy. Spores cream, subglobose, 6–9 × 6–8 µm. Taste mild. Edible and good.

1 *Russula claroflava* (Chrome Yellow Russula, Clear Yellow Russula): *in boggy birch woods; cap clear bright yellow, gills yellowish; flesh slowly blackish*

Autumn. Frequent. Cap 5–10 cm (2–4 in), convex then expanded, shining and moist at first then soon dry and smooth: margin often slightly striate; brilliant chrome to lemon-yellow, whole fungus slowly (5–10 hours) greying to black. Gills pale primrose yellow. Stem rather tall, equal, soft, white. Spores 8–10 × 7–8 µm, ovate, in deposit rather deep cream. Taste mild to slightly hot, especially when young. Edible and delicious. Frequently confused with *R. ochroleuca* (see below) but distinguished by its habitat, much brighter yellow cap and yellowish gills; *R. ochroleuca* does not go grey-black.

2 *Russula caerulea* (Violet Russula): *below pines; cap umbonate, deep violet, bluish purple to brownish purple*

Autumn. Occasional. Cap 4–8 cm (1½–3 in), almost conical then expanded with blunt umbo (this is almost unique within the genus). Gills crowded, pale ochre. Stem rather long, slender, firm, white. Spores in deposit pale ochre, ovate, 8–10 × 7–9 µm. Taste of flesh and gills mild, cuticle bitter. Edible but not recommended.

3 *Russula emetica* (Sickener): *below pines; cap bright scarlet; white gills and stem; very hot*

Usually in damp, boggy ground. Autumn. Common. Cap 5–10 cm (2–4 in), convex then expanded, bright clear scarlet or blood-red, sometimes paling with age or after rain: cuticle peels completely; margin soon sulcate. Gills pure white to pale cream. Stem slightly bulbous, soft, fragile, pure white. Spores white in deposit, ovate, 9–11 × 7–9 µm. Taste extremely hot. Can cause vomiting when raw, apparently harmless when cooked; best avoided. The very similar *R. mairei*, also scarlet, grows in beech woods only, has the cuticle only half peeling, white gills with greenish grey reflections, and odour of honey or fruit.

4 *Russula ochroleuca* (Common Ochre Russula): *in any woods; cap dull ochre to greenish yellow*

Summer and autumn. Abundant. Cap 4–10 cm (2–4 in), convex then flattened; margin sulcate with age. Gills white to pale cream, brittle. Stem soft, often soon hollow, slightly striate with age; white then greyish (not blackening). Spores off-white in deposit, ovate, 8–10 × 7–8 µm. Taste mild to slightly hot.

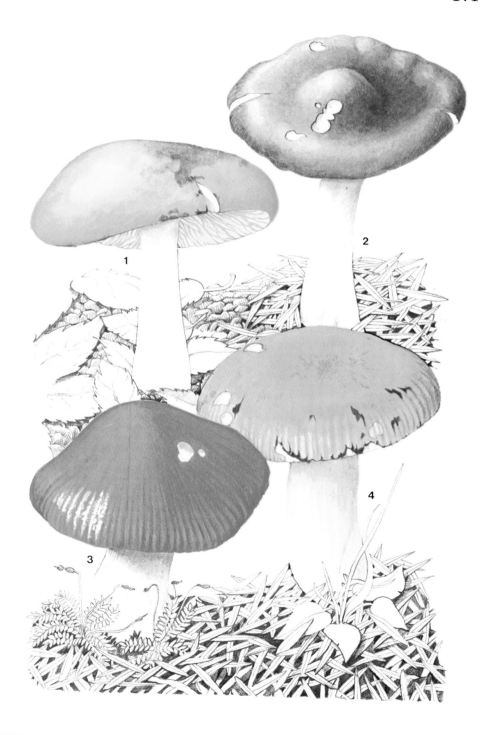

RUSSULA (contd)

Edible but not recommended. Easily recognized by the dull ochre yellows of the cap and the pale gills. This is one of the commonest of all toadstools.

1 *Russula foetens* (Stinking Russula, Foetid Russula): *cap large, viscid at first; unpleasant odour and taste*

In mixed woods. Summer and autumn. Common. Cap 5–15 cm (2–6 in), robust, very firm, convex then flattened; glutinous-viscid then dry, with brown flocci or scabs at margin, often strongly furrowed and striate; dull yellowish to honey-brown. Gills thick, distant; cream often discoloured with rust-like spots. Stem stout, firm, hollow, brittle; whitish to dull yellowish. Odour strong, oily or rancid. Spores in deposit pale cream, subglobose, 8–10 × 7–9 µm. Taste of gills very hot, of stem almost mild. Not edible. The related *R. laurocerasi* is most easily distinguished by its odour of bitter almonds (marzipan). *R. illota*, uncommon, has a similar odour (less distinct, often also unpleasant), cap coated with violaceous gluten and gill edges minutely spotted violet-brown (lens required).

2 *Russula delica*: *white to cream, rather funnel-shaped; white gills and stem apex often bluish*

In mixed woods. Summer and autumn. Common. Cap 5–15 cm (2–6 in), convex, soon expanded, usually funnel-shaped; rather velvety at first especially at strongly inrolled margin. Gills crowded, rather decurrent. Stem rather short, firm. Spores white, ovate, 8–12 × 7–9 µm. Taste hot and bitter. Odour rather unpleasant. Edible but not recommended.

3 *Russula fellea* (Geranium-scented Russula): *overall straw-honey colour; odour of geranium*

In deciduous woods. Autumn. Common. Cap 4–8 cm (1½–3 in), convex then expanded, fleshy, firm, margin slightly sulcate; ochre-yellow to straw but less dull than in *R. ochroleuca* (previous page). Gills pale straw-yellow to honey. Stem rather stout, firm, colour as cap and gills. Spores in deposit pale cream, ovate, 7–9 × 6–7 µm. Odour of household geranium (*Pelargonium*) although often faint and not detectable by some. Taste very hot. Edible when cooked but best avoided.

4 *Russula fragilis* (Toothed-gill Russula): *small, fragile, colours purplish (but variable); toothed gill-edge*

In mixed woods. Summer and autumn. Very common. Cap 2–6 cm (1–2½ in), convex then expanded and often depressed, very soft and fragile, margin furrowed: mixtures of purple, greenish, violet and black all common, centre usually darker. Gills white to pale cream, margin distinctly toothed when viewed under lens. Stem white, soft, fragile (difficult to pick intact). Spores white in deposit, subglobose, 7–9 × 6–8 µm. Taste very hot. Edible when cooked but best avoided. Even if it is washed out and almost white, the gill edge is distinctive.

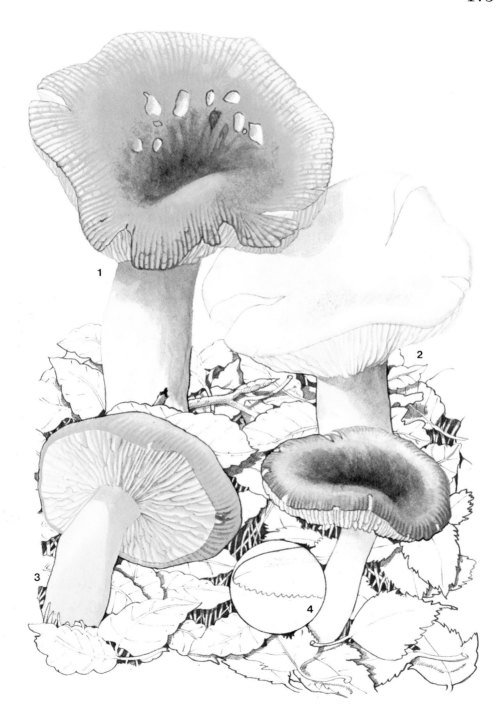

RUSSULA (contd)

1 Russula cyanoxantha: *violet-green colours of cap; gills soft, flexible; specific chemical test*

In mixed woods. Summer and autumn. Common. Cap 5–15 cm (2–6 in), convex then expanding, sometimes depressed, firm-fleshed; colours very variable, usually some shade of violet to bluish-green, often both in one cap. Gills narrow, fairly crowded, white; soft and greasy, "elastic" to the touch, not breaking. Stem rather stout, firm; white or with faint violet flushes. Spores white, elliptic, 7–9 × 6–7 µm. Taste mild. Chemical test: iron sulphate ($FeSO_4$) + flesh = no reaction to *faintly* green. Edible and recommended. The closely related *R. ionochlora*, also edible, more common, has brittle, darker cream gills, spores cream, and flesh + iron sulphate = salmon pink.

2 Russula lepida: *very firm flesh; stem and cap deep rose-red to pink with white "bloom"; cedarwood taste*

In beech woods. Autumn. Frequent. Cap 4–10 cm (1½–4 in), convex then flattened, very firm-fleshed; sometimes fading to yellowish or white. Cuticle hardly peeling. Gills crowded, narrow, off-white to cream, brittle. Stem equal, very firm to hard-fleshed, white or more often flushed as cap. Spores white, sub-globose, 8–9 × 78 µm. Taste pleasant, odour slightly fruity, of menthol or distinctive of cedarwood, as in a pencil. Edible but not recommended. The closely related *R. amarissima*, uncommon, has a bitter taste and rather more purplish red cap. *R. rosea*, common in deciduous woods, differs in the cap not pruinose, flesh not so hard, stem rarely pink, cuticle peeling halfway to the cap centre, and without distinctive odour or taste.

3 Russula lutea: *small, soft, fragile toadstool; cap yellow to pink; gills deep ochre-orange*

In deciduous woods. Summer and autumn. Common. Cap 2–5 cm (1–2 in), convex then expanded, often depressed; margin often furrowed, cuticle almost completely peeling; apricot-yellow, peach or coral to entirely clear rose-pink, almost white with age or after rain. Gills rather broad, strongly interveined. Stem white. Spores deep ochre-orange, ovate 7–9 × 6–8 µm. Taste mild. Odour when fresh of apricot. Edible but rather too small and soft. The related *R. nauseosa*, under pines, has a duller, more purple-brown cap and slightly hot taste.

4 Russula queletii: *purple cap and stem (especially at base); gills cream; no reaction in ammonia test*

Below conifers. Autumn. Occasional. Cap 4–10 cm (1½–4 in), convex then expanded, deep purple-red, red or purple-violet, firm, fleshy. Gills very slightly adnate-decurrent. Stem usually slightly swollen at base; pruinose. Spores pale cream-ochre, subglobose, 8–10 × 7–9 µm. Taste hot. Not edible, bitter when cooked. The very similar *R. sardonia* has distinctly yellowish gills, and gills + ammonia = deep rose-red in 15 minutes.

RUSSULA (contd)

1 Russula nigricans (Blackening Russula): *large "solid" fungus, turning brown-black all over; gills widely spaced*

In mixed woods. Summer and autumn. Abundant. Cap 6–20 cm (2½–8 in), convex then expanded and depressed, very firm, solid fleshed; white then soon brown and finally coal-black as if burned. Gills very thick, broad, brittle, widely spaced, cream. Stem short, stout, firm, white. Flesh white, soon red, finally black. Spores white, ovate 7–8 × 6–7 μm. Taste slowly hot. Edible but not recommended. *R. densifolia* is extremely similar but differs in its crowded gills. *R. albonigra*, uncommon, has gills quite crowded and flesh turning directly black, no red stage.

2 Russula virescens (Green Cracking Russula): *cap cracking into little polygonal platelets, usually green*

In beech woods. Summer and autumn. Frequent. Cap 5–10 cm (2–4 in), subglobose then convex and expanding, very fleshy, firm; pale verdigris to herbage green, often paling to yellow or white. Gills crowded, brittle, white to cream. Stem rather short, stout, firm; white, slightly staining brown. Spores white, subglobose, 7–9 × 6–7 μm. Taste mild. Edible and delicious.

3 Russula vesca (Bare-edged Russula): *cap pinky-brown to buff; cuticle retracting from margin, showing flesh below*

In deciduous woods. Summer and autumn. Common. Cap 5–10 cm (2–4 in), convex then expanded, fleshy, firm; colour very variable, usually shades of pale reddish brown, buff, pinkish brown to flesh, olive or almost white; cuticle will also peel to half way. Gills narrow, crowded, forking near stem, white to pale cream. Stem firm, white, base often pointed, stained rust-brown. Spores white, ovate, 6–8 × 5–6 μm. Taste mild. Chemical test: iron sulphate ($FeSO_4$) on flesh = rapidly extremely deep salmon pink. Edible and good.

4 Russula xerampelina: *very variable, but the unique iron sulphate reaction (dark green) is constant*

In mixed woods. Summer and autumn. Common. Cap 5–15 cm (2–6 in), convex then expanded, fleshy, firm; colour extraordinarily variable, mixtures of brown, green, red and purple or even yellow-ochre, usually rather dull-looking; surface distinctly matt, colours often slightly "zoned". Gills broad, thick, pale cream to ochre. Stem usually rather stout, white or flushed reddish brown, surface rather "veined". Spores pale ochre, ovate, 8–11 × 6–9 μm. Odour, particularly of crushed gills, often strong of crab, fish (herring), shrimps. Taste mild. Chemical test: iron sulphate on flesh = deep green; this reaction makes an otherwise almost impossible "chameleon" easy to confirm. Edible but not recommended.

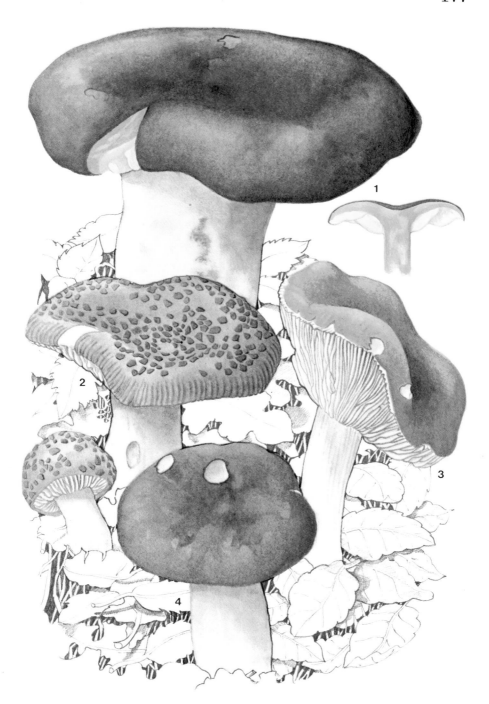

LACTARIUS

1 *Lactarius blennius* (Slimy Milk-cap): *cap slimy, greenish, marked with darker spots; very hot milk*

In beech and oak woods. Autumn. Frequent. Cap 4–8 cm (1½–3 in), convex then rather flat, very slimy at first, especially after rain; pale greenish grey to olive-brown, slightly zoned and with depressed spots usually darker. Gills somewhat adnate, white to dull grey. Milk white, fairly copious, slowly turns grey, extremely hot to taste. Stem rather short, stout, colour as cap, viscid. Spores cream, ovate, 7–8 × 6–7 μm. Not edible. The rather rare *L. circellatus*, found under hornbeam, has a greyer, distinctly zonate cap, often very irregular in outline, and a very short stem.

2 *Lactarius helvus:* *cap pale cinnamon, minutely squamulose; smell sweet, spicy; milk mild*

In damp birch woods. Autumn. Occasional. Cap 5–10 cm (2–3 in), convex then soon flattened and often depressed to funnel-shaped; finely felty-squamulose, pale ochre, cinnamon-brown to yellow-brick. Gills slightly decurrent, thin, crowded; watery, yellowish-flesh to cinnamon. Milk thin, scanty, mild. Stem rather soft, long, colour as cap. Spores pale ochre, subglobose, 7–9 × 5–6 μm. Edible but not recommended.

3 *Lactarius chrysorrheus* (Yellow Milk-cap): *cap yellowish-pink; milk white, turning bright yellow*

In oak woods. Autumn. Frequent. Cap 4–8 cm (1½–3 in), convex then flat, rather viscid at first; pinkish flesh-colour to yellowish with darker zones. Gills slightly decurrent, fairly crowded, cream. Milk copious, white then bright sulphur-yellow, very bitter to taste. Stem firm, rather short, colour as cap. Spores white, subglobose, 6–7 × 6 μm. Not edible.

4 *Lactarius quietus* (Oily Milk-cap): *only under oak; cap brown, zoned; characteristic oily, sweet or rancid odour*

Summer and autumn. Abundant. Cap 4–8 (1½–3 in), convex then expanded, dry, smooth; dull reddish brown with darker concentric zones. Gills pale whitish to pale brown, adnate-decurrent, fairly crowded. Milk white, mild and sweet. Stem rather tall, soft, colour as cap. Spores cream, 7–9 × 6–7 μm. Edible but not very good.

5 *Lactarius rufus* (Red Milk Cap): *cap foxy-red, minutely umbonate; milk white, mild then slowly very hot*

In birch or pine woods. Autumn. Very common. Cap 4–8 cm (1½–3 in), convex then soon expanded and often depressed, almost always with small sharp umbo; rich reddish "foxy" brown, texture slightly roughened. Gills adnate-decurrent, reddish ochre. Stem hollow, colour as cap. Spores cream, ovate, 8–9 × 6–7 μm. Not edible.

LACTARIUS (contd)

1 *Lactarius torminosus* (Woolly Milk Cap): *cap pinkish, shaggy and zoned; milk acrid*

In damp birch woods. Autumn. Frequent. Cap 4–12 cm (1½–5 in), convex then expanded, depressed; margin at first strongly inrolled, and extremely woolly-shaggy; pale pinkish to yellowish flesh-colour with faint zones. Gills adnate-decurrent, thin, crowded, paler than cap. Milk white, copious, extremely hot. Stem rather short, colour as cap. Spores cream, ovate, 7–9 × 6–8 µm. Not edible. The less common *L. pubescens*, under birch, has a paler, unzoned cap and smaller spores.

2 *Lactarius deterrimus* (Saffron Milk Cap): *orange with concentric zones, staining green; copious carrot-red milk turning deep green; flesh orange then red, finally green*

In conifer woods. Autumn. Common. Cap 4–10 cm (1½–4 in), convex then expanded and slightly depressed, viscid when moist then dry, orange-brick to reddish with concentric zones staining greenish with age or bruising. Gills adnate-decurrent, crowded, orange-ochre. Stem rather short, solid, colour as cap. Spores cream, 7–9 × 6–7 µm. Edible but not particularly good. This species has for many years been known in Britain as *L. deliciosus*, which is in fact known only from mainland Northern Europe and, rarely, northern Britain, differing in its pleasant taste, and flesh turning directly green without the red stage.

3 *L. tabidus*: *cap with tiny umbo; milk turns yellow on handkerchief*

In deciduous woods. Autumn. Common. Cap 3–5 cm (1–2 in), convex then expanded and usually with a minute umbo; surface dry, slightly roughened and puckered at centre; pale brick-red to tan, paler when dry. Gills adnate-decurrent, slightly distant, colour as cap. Milk white, turning yellow in about one minute, taste mild. Stem rather soft, slender, tapered above, colour as cap. Spores cream, ovate, 8–10 × 5–7 µm. Edible but not recommended.

The similar *L. subdulcis* differs in its slightly pinker cap without an umbo, and white milk not turning yellow.

4 *Lactarius vietus*: *cap grey-lilac; milk turns deep grey*

In damp deciduous woods especially birch. Autumn. Frequent. Cap 4–10 cm (1½–4 in), convex then expanded and shallowly depressed; smooth, slightly viscid when moist; pale greyish brown with a flush of lilac. Gills adnate-decurrent, thin, crowded, white to pale ochre. Milk white, turning grey on the gills in about 20 minutes; taste hot. Stem rather soft, slightly paler than cap. Spores pale cream, ovate, 8–9 × 6–7 µm. Not edible.

L. uvidus has flesh when cut turning bluish lilac and milk white, mild and unchanging.

LACTARIUS (contd)

1 _Lactarius piperatus_ (Peppery Milk-cap): _cap white, smooth when young; gills crowded; milk hot_

In deciduous woods. Summer and autumn. Common. Cap 5–15 cm (2–6 in), convex, soon expanded and depressed; white to pale cream, smooth, cracking into plates with age; margin inrolled at first. Gills adnate-decurrent, crowded, thin, repeatedly forking, cream to yellowish. Milk white, copious, very hot and peppery. Stem firm, solid, stout, tapering downwards, colour as cap. Spores white, ovate, 6–9 × 5–7 μm. Not edible. The similar and slightly larger (to 25 cm) _L. vellereus_ also has a depressed cap, but has a woolly-tomentose surface, especially at the strongly inrolled margin, gills rather widely spaced, and a short stout stem.

2 _Lactarius turpis_ (Ugly Milk-cap): _ugly, dark cap, olive-brown to black; gills yellowish; milk hot; specific chemical test_

In deciduous woods on damp, boggy soil. Autumn. Common. Cap 5–15 cm (2–6 in), convex then expanded and slightly depressed, margin often rather irregular, inrolled; viscid when moist, colour very dull, greenish brown to almost sepia or black, more yellowish at margin. Gills adnate-decurrent, crowded, dull straw-yellow, spotted reddish brown. Milk white, copious, very hot and unpleasant. Stem rather short, viscid, surface often pitted and spotted, deep olive. Spores cream, ovate, 7–8 × 6–7 μm. Chemical test: ammonia on any part = deep violet. Not edible.

3 _Lactarius volemus_: _rich tawny orange to reddish orange cap, (often cracking) and stem; smell of herring_

In deciduous woods. Autumn. Uncommon. Cap 5–10 cm (2–4 in), convex then expanded and slightly depressed; surface dry, matt, often minutely cracking concentrically. Gills adnate-decurrent, rather crowded, cream. Milk white, turning slightly brownish, copious, mild, odour of herrings. Stem often tall; stout, firm, slightly hollow, smooth. Spores white, globose, 8–10 μm. Edible but not very good.

In North America the very similar _L. hygrophoroides_ is also found, differing in its widely spaced gills.

4 _Lactarius glyciosmus_ (Coconut Milk-cap): _cap greyish-lilac with umbo; smell of dried coconut_

In deciduous woods, especially birch. Autumn. Frequent. Cap 3–6 cm (1–2½ in), convex then expanded with obtuse umbo; pale greyish lilac, dry. Gills adnate-decurrent, thin, yellowish flesh-colour. Milk white, mild at first then slightly hot. Stem soft, slightly longer than cap diameter, colour as cap only paler. Spores deep cream, ovate, 7–8 × 6–7 μm. Edible and quite good. Also with a coconut odour, but under conifers, is _L. mammosus_, cap darker, brownish, with acute umbo; rather rare.

BOLETACEAE

The Boletes are found throughout the world in both temperate and tropical climes and without doubt can claim some of the most luridly coloured, delicious and often huge fruit-bodies in the entire range of agarics. There are species whose caps reach two feet (60 cm) in diameter with gross stems to match, while others are only a few centimetres across. Almost all share the characteristics of a soft, fleshy body, and spores produced in a tubular fertile layer (hymenium) instead of on gills. These tubes are densely packed under the cap, their exits (the only part visible without breaking or cutting the cap) being referred to as pores. Colour changes within the flesh are extremely common within this family and are valuable guides to identification. Almost all species are considered edible, with only a few exceptions which are either just bitter or mildly upsetting. One species, the Cep (*Boletus edulis*) is one of the most famous edible fungi in the world.

Genera included within the family and described here are: *Boletus, Leccinum, Suillus, Aureoboletus, Boletinus, Strobilomyces, Gyroporus, Porphyrellus* and *Tylopilus*. Many of these were formerly included in the genus *Boletus* but are now believed to be genera in their own right. Included within the wider order Boletales are some toadstools which, despite having gills, not pores, are believed to be closer to the rest of the boletes than to other gilled fungi; these are the Gomphidiaceae and the Paxillaceae, pp. 202–5.

BOLETUS

1   ***Boletus satanus*** (Devil's Bolete, Satan's Bolete): *large cap pale, hardly flushed red at margin, if at all; red network on lower stem; flesh very pale yellow then bluish*

In deciduous woods especially on chalky soils. Late summer and early autumn. Uncommon. Cap 10–30 cm (4–12 in), thick-fleshed, pale greyish white to greenish grey; smooth, minutely tomentose especially when young; often flushed red at margin. Tubes yellowish green, blue on cutting; pores small, round, blood-red to orange with age, yellow at the extreme margin, green-blue upon bruising. Stem very stout, clavate, orange to yellow at apex flushing red to purplish in lower half, with raised red network over most of stem. Flesh pale yellow becoming pale bluish on cutting. Spores olive-brown, subfusiform, 11–14 × 4–6 μm. *Reputedly very poisonous* but probably just upsetting and emetic.

The related *B. purpureus* is distinguished by its very pale cap flushing pink-purple and bruising blue, and flesh bright yellow turning blue.

BOLETUS (contd)

1 *Boletus aereus:* cap very dark sepia; stem rich brown, with network

In deciduous woods. Summer and autumn. Uncommon, mainly southern distribution. Very similar to *B. edulis* but cap distinctly rough, granulate, tending to crack into minute scales; colour deep intense sepia-brown, almost black, often with discoloured, paler patches; and stem strongly coloured, rich reddish brown with brown network. Edible and delicious: a beautiful and distinctive species. *B. aestivalis* has a similar cap texture, but colour pale, ochre-brown to cinnamon; stem buff with dense white net.

2 *Boletus edulis* (Cep, Penny Bun): *cap rich brown, robust; stem paler with white network; flesh firm, white, almost unchanging*

In deciduous or conifer woods. Summer and autumn. Common. Cap 5–18 cm (2–7 in), convex then expanded, smooth, slightly greasy in wet weather; thick-fleshed and firm; rich toasted brown, chestnut or bay; margin paler, white at edge. Tubes white then yellowish. Pores white then cream, finally lemon-yellow in some forms. Stem robust, often very fat (tall and straight in subspecies *trisporus*, with three-spored basidia); paler than cap, almost white, with a white network, especially at apex. Spores olive-brown, subfusiform, 14–17 × 4–6 µm. Edible and perhaps the most delicious of all fungi. The related *B. pinophilus* occurs only under conifers; cap and stem are wine-red with vinaceous tints; flesh also turning vinaceous.

3 *Boletus calopus:* cap greyish clay; pores yellow; stem pale to deep red with fine white network overall; flesh bitter

In deciduous and conifer woods. Autumn. Uncommon. Cap 5–15 cm (2–5 in), convex, thick-fleshed, margin inrolled when young; smooth, dry, pale greyish clay or almost white, darker with age. Tubes and pores dull, pale yellow, bluish when cut. Stem stout, often swollen, firm and solid; apex yellow. Flesh cream to lemon-yellow, blue when cut. Spores olive-brown, subfusiform, 12–16 × 4–6 µm. Not edible. The related *B. albidus* differs in the paler whitish cap, often very large (20 cm, 8 in); stout stem, pale yellow without or with only very slight network; taste also bitter.

4 *Boletus badius* (Bay-capped Bolete): *bay-coloured cap and stem; pores pale yellowish, pale blue on bruising and cutting*

In deciduous and conifer woods. Autumn. Abundant. Cap 4–15 cm (1½–6 in), convex, at first minutely tomentose then smooth and often slightly viscid when wet; deep bay-brown to rather pale ochre-brown in some forms. Tubes cream to pale yellow, turning grey to bluish green. Stem equal, straight, solid, ochre to pale chestnut, smooth. Flesh pale, white to yellowish, turning bluish on cutting. Spores olive-brown, subfusiform, 13–15 × 4–6 µm. Edible and quite good. There is a

BOLETUS (contd)

form found in beech woods, considered by some a valid species (*B. vaccinus*), with cap chestnut, tomentose, and white unchanging flesh.

1 Boletus erythropus: *cap deep brown; pores deep red; red stem without network; yellow flesh instantly deep blue if cut*

In deciduous and conifer woods. Summer and autumn. Common. Cap 5–10 cm (2–4 in), convex then expanded, dry, smooth to slightly tomentose; rich chestnut- to olive-brown, yellower at margin. Tubes lemon-yellow, bluish green on cutting. Pores sometimes almost maroon, paler, more orange when expanded, margin usually yellow. Stem stout, smooth, without network, under lens visible as red stippling on yellow. Spores olive-brown, subfusiform, 12–15 × 4–6 μm. Edible but not recommended. Commonest of the red-pored, blue-staining boletes.

2 Boletus luridus: *yellowish, red-flushed stem with prominent red network; flesh blueing; pores orange*

In deciduous woods especially on calcareous soils. Autumn. Occasional. Cap 5–15 cm (2–6 in), convex then flattened, minutely tomentose then smooth; very variable from dark flesh-pink to olivaceous and brown, usually retaining pink at margin and rather rust-coloured at centre; bruising blue-black. Tubes yellow-green, blue on cutting. Pores orange-red to pale orange bruising deep blue; usually yellow at margin. Flesh yellowish in cap and upper stem, brighter yellow below, often purplish red in base; turning blue; *usually* with red line of flesh above tube-layer. Spores olive-brown, subfusiform, 11–15 × 4–7 μm. Apparently edible when cooked, but best avoided. The similar but rarer *B. queletii* has a rich orange to reddish brown cap, orange pores and stem yellowish above, purple-red below as is flesh in stem; no network.

3 Boletus appendiculatus: *cap brick-red to rust-brown; bright yellow pores; overall yellowish network on stem*

In deciduous woods. Late summer, early autumn. Uncommon, mainly southern. Cap 5–15 cm (2–6 in), convex then slightly expanded, dry, slightly roughened; rich brick-red, bay to rust, often cracking. Tubes lemon-yellow, blue when cut. Pores bright lemon to golden yellow, intense blue when bruised. Flesh pale yellow, blue and patchily pink on cutting. Stem robust, yellow above becoming reddish below, often rust-spotted at base, with net; bruising blue-green. Spores olive-brown, subfusiform, 12–15 × 3–5 μm. Edible and good. *B. fechtneri*, closely related, has a very pale greyish buff cap bruising coffee-brown, stem yellow with red zone. The rare *B. regius* is very similar but has a rich carmine-red cap.

4 Boletus porosporus: *cap dull sepia; cracking; spores with a distinct pore*

In mixed woods. Autumn. Frequent. Cap 4–8 cm (1½–3 in), con-

BOLETUS (contd)

vex then flattened, tomentose, soon cracking to expose whitish flesh beneath; olive-brown, soon dull, dirty sepia. Tubes yellow-olive. Pores angular, large, lemon-yellow then olivaceous, finally dull brownish. Flesh pale lemon-yellow in cap and upper stem, reddish brown below, turning bluish when cut. Stem fairly slender, tall; apex yellowish, then a narrow band of red, dull greyish sepia below; no network but some longitudinal ridges. Spores olive-brown, subfusiform, truncate at one end (and with a distinct pore), 13–15 × 4–6 µm. Edible but not recommended.

1 *Boletus subtomentosus* (Downy Bolete): *woolly cap, yellow-green to brown; any cracks expose pale, not red, flesh*

In mixed woods. Summer and autumn. Common. Cap 5–12 cm (2–5 in), convex then expanded, tomentose, hazel-brown to olivaceous-buff; woolly tomentum often bright yellowish green. Pores and tubes lemon to chrome-yellow bruising only slightly blue if at all. Stem rather slender, buff to reddish brown or olive, often with slight ridges at apex. Flesh pale yellow to brownish, date-brown below cuticle. Spores olive-brown, subfusiform, 10–13 × 3–5 µm. Edible but not very good. Chemical test: ammonia on cap cuticle = immediately date-purple. The similar *B. lanatus* has strong brown ribs almost as a net on stem apex, pores yellow turning blue, and ammonia making cap blue-green, then fading.

2 *Boletus chrysenteron* (Red-cracked Bolete): *cap cracking to expose reddish flesh*

In mixed woods. Autumn. Abundant. Cap 4–12 cm (1½–5 in), convex then flattened, subtomentose then smooth; hazel to olivaceous brown or buff, often a purplish flush in colder weather and ceasing to crack. Tubes lemon-yellow then greenish. Pores large and angular, lemon-yellow then greenish, sometimes bruising blue. Flesh cream-yellow to lemon, browner in stem base, slightly blue when cut. Stem rather slender, smooth, yellowish to buff with reddish floccose granules; very variable. Spores olive-brown, subfusiform, 12–15 × 3–5 µm. Edible but not very good. A rare species, with an apricot-coloured cap is *B. armeniacus*, found in Europe and Britain.

3 *Boletus pruinatus:* *cap dark, reddish bay to black with white "bloom" when young; chrome-yellow stem and flesh*

In beech and oak woods. Late autumn. Occasional to common. Cap 3–6 cm (1–2½ in), convex. Tubes, pores lemon-yellow, bruising bluish. Flesh bruising bluish. Stem rather swollen, smooth, reddish at base, noticeable yellowish mycelium matting leaf-litter. Spores olive-brown, subfusiform, 11–14 × 4–6 µm. Edible and quite good. Beautiful, distinct features but often overlooked, sometimes confused with others.

(4, see over)

BOLETUS (contd)

4　Boletus versicolor (Red-capped Bolete): *carmine-red to purplish cap and stem; all parts bruise blue*

In grass verges near oaks. Autumn. Frequent. Cap 3–8 cm (1–3 in), convex then expanded; subtomentose, sometimes cracking, soon fades to reddish brown or olive. Pores and tubes lemon-yellow then greenish. Flesh dull buff-yellow, reddish in stem. Stem rather slender, tapered at base, blood-red to carmine or purplish, yellow at apex. Spores olive-brown, subfusiform, 11–14 × 4–6 μm. Edible but not very good.

1　Boletus parasiticus: *immediately recognizable since it grows only on earthballs*

Always on the common earthball (*Scleroderma citrinum*). Autumn. Uncommon. Cap 2–4 cm (1–1½ in), semiglobate to convex, hardly expanding; rich ochre-yellow to olivaceous brown; surface velvety, sometimes cracking. Tubes and pores bright lemon-yellow, pores often stained reddish. Stem rather stout, short, tapered at base, which attaches to the bottom of the earthball, colour as cap. Spores olive-brown, fusiform, 11–21 × 3–5 μm. Edible but not recommended.

2　Boletus piperatus (Peppery Bolete): *cinnamon-coloured cap and stem; rust-coloured pores; bright yellow stem base; peppery taste*

Usually under birches, often in association with the Fly Agaric (*Amanita muscaria*, pp. 52–3). Autumn. Common. Cap 3–10 cm (1–4 in), convex then expanding, smooth to slightly greasy-viscid when moist; ochre-brown to cinnamon. Tubes and pores cinnamon to rust, pores large and angular. Stem rather slender, colour as cap except base which is bright golden yellow, as is flesh in base. Flesh in rest of stem and cap pale buff-ochre to reddish. Spores olive-brown with cinnamon tints, subfusiform, 8–11 × 3–4 μm. Not poisonous but usually considered inedible because of very peppery taste.

LECCINUM

3　Leccinum scabrum (=*Boletus scabrus*) (Brown Birch Bolete): *cap nut-brown, very soft; stem scabrous; flesh white, almost unchanging*

Under birches. Summer and autumn. Common. Cap 5–15 cm (2–6 in), convex then expanding, smooth, moist and sticky in wet weather, otherwise dry; pale hazel-brown to buff. Tubes and pores very pale, almost white, then clay-buff to ochre with age, bruising cinnamon. Stem rather long, white to greyish buff covered with small, woolly, brown to blackish scales (scabrosities). Flesh very soft especially in cap, white and unchangeable to very slightly pale pinkish buff; sometimes vivid blue-green or yellow at extreme base of stem. Spores tobacco-brown, subfusiform, 14–20 × 5–6 μm. Edible and quite good.

This species has now been split into several closely related species among which are *L. roseofracta*, stouter and darker brown with very dark, blackish stem scabrosities and flesh

LECCINUM (contd)

strongly pink or reddish when cut, and *L. variecolor* with grey-ish, marbled cap and greyish scabrosities, flesh bright pink in cap and upper stem, bright blue-green below.

1 Leccinum quercinum: *cap deep fox-red to chestnut; stem scales reddish brown; under oaks*

Late summer and autumn. Frequent in southern England and Europe. Very similar in all respects to the more common *L. versipelle* (below) but distinguished by its habitat, the deep fox-red to orange-chestnut cap and stem scabrosities, and pores not dark grey when young. Spores brown, fusiform, 13–18 × 4–5 µm. This is a much confused species which has only recently been sorted out. Edible and good.

2 Leccinum versipelle (=Boletus testaceoscaber): *cap often tawny-orange; pores dark when young; stem with blackish scales*

Always under birch. Summer and autumn. Common. Cap 6–20 cm (2½–8 in), convex then expanding, with narrow pendent fringe of cap cuticle at margin; surface dry, downy to smooth with slightly adpressed-scaly centre; bright yellow-orange to orange-buff or tawny. Tubes white to buff, pores small, very dark greyish to black when young then soon paler, buff. Stem rather stout, white to buff, densely speckled with darker brown to black woolly scabrosities; often bruising greenish at base when handled. Flesh white, then pinkish, finally violaceous grey to black when cut. Spores tobacco-brown, subfusiform, 12–16 × 4–5 µm. Edible and highly regarded by many.

L. aurantiacum differs in growing below aspens and in its richer orange cap and orange stem scales, pores not dark at first. See also the previous species.

3 Leccinum holopus: *white cap, pores, stem and stem scabrosities; cap greenish buff with age; in sphagnum bogs under birch*

Especially in northern Europe, rarely in southern districts. Autumn. Occasional. Cap 4–12 cm (1½–5 in), convex then expanding, downy when young, smooth and viscid with age; pure white then pale buff with greenish tints. Pores white to buff. Stem rather slender. Flesh white, unchanging. Spores tobacco-brown, subfusiform, 15–18 × 5–6 µm. Edible and quite good.

4 Leccinum crocipodium (Yellow Cracking Bolete): *cap bright yellow, then ochre-brown cracking; yellow pores and scaly stem; all parts black when cut or bruised*

Below oaks, in Britain only in southern parts. Late summer and early autumn. Uncommon. Cap 4–10 cm (1½–4 in), convex then expanded, smooth and slightly downy when young; bright lemon-yellow to yellow-orange when fresh, soon dulling to ochre-brown, cracking all over like mosaic with age to show

white flesh. Tubes and pores lemon-yellow. Stem rather stout, pale yellow with scabrosities of same colour. Spores ochre-brown, subfusiform, 12–17 × 4–7 μm. Edible and good.

1 *Suillus grevillei* (=*Boletus elegans*) (Larch Bolete): *only under larch; cap yellow to brick-colour; yellow stem with ring; pores lemon-yellow*

Late summer and autumn. Common. Cap 4–10 cm (1½–4 in), convex then expanded, very viscid when moist, smooth and polished when dry; darker with age and distinctly dark brick-colour in the variety *badius*. Tubes rather decurrent, pale yellow. Pores quite small, angular, lemon-yellow, bruising rust-colour. Stem rather slender to medium, with a distinct fleshy ring; pale yellow above, more ochre below with reddish stains. Flesh yellowish becoming slightly blue in stem base. Spores ochre, subfusiform, 8–11 × 3–4 μm. Edible but not very good.

2 *Suillus* (=*Boletus*) *luteus* (Slippery Jack): *under pines; cap deep purplish brown to chestnut; ring, with violet flush below*

Below pines. Autumn. Common. Cap 5–10 cm (2–4 in), convex then expanded and slightly umbonate, smooth and very viscid-glutinous, shiny when dry; deep purplish brown to chocolate-brown or chestnut. Tubes pale yellow to straw-colour. Pores straw-yellow to yellow-olive. Stem medium pale straw with large white or violaceous fleshy ring above, flushed violaceous below ring, apex with darker glandular dots. Flesh yellowish. Spores ochre, subfusiform, 7–10 × 3–4 μm. Edible but not highly recommended.

3 *Suillus aeruginascens* (=*Boletus viscidus*): *pallid buff-olive to greyish colours; whitish ring; under larch*

Autumn. Rather uncommon. Cap 4–10 cm (1½–4 in), convex then expanded, rather wrinkled and irregular, very viscid when moist; pale olive-buff to cream or straw, often blotched with greyish or brownish spots. Tubes and pores pale olive-buff to straw with grey-green flush. Stem rather short, slender, colour as cap but with greyish flush at apex; white or greyish ring above. Spores tobacco brown, subfusiform, 10–12 × 4–5 μm. Edible but not recommended. This species is very distinctive because of its colouring.

SUILLUS (contd)

1 Suillus (=Boletus) **bovinus:** *cap orange-buff with white margin; no ring*

Below pines. Autumn. Common. Cap 4–10 cm (1½–4 in), convex then expanded, smooth and viscid, pale buff to clay with an ochre tint. Tubes rather decurrent, dull olive-grey. Pores rather large and angular, "compound" (i.e. there are pores within pores), dull yellowish olive to buff or ochre. Stem rather short, tapered, colour as cap. Spores olive-brown, subfusiform, 8–10 × 3–4 μm. Edible and quite good.

2 Suillus variegatus: *tawny-ochre colours and small felty scales on cap; no ring; cap only slightly moist*

Under conifers, especially pines, and on poor sandy soils. Autumn. Common. Cap 4–10 cm (1½–4 in), convex then slightly expanded, rather thick-fleshed, slightly moist but usually subtomentose especially when young; tawny ochre to sienna or rust with small, slightly darker, felty scales. Tubes and pores yellowish, distinctly flushed olive at first, finally cinnamon, bruising greenish. Stem rather short, stout, often swollen, colour as cap but matted paler; base often with a matted, rooting mycelial mass spreading outwards. Flesh pale yellowish often flushing pale blue when cut. Spores tobacco-brown, subfusiform, 9–11 × 3–4 μm. Edible and quite good.

3 Suillus granulatus: *cap reddish brown to golden ochre, smooth and viscid; stem (without ring) and pores exude milky droplets in damp weather*

Below pines. Autumn. Common. Cap 3–8 cm (1–3 in), convex then expanded; reddish fawn to rust, finally often golden ochre. Tubes and pores pale lemon-yellow. Stem rather slender, pale yellowish with pinkish or coral base; apex with white or yellowish floccose granules, which also exude droplets. Spores ochre, subfusiform, 8–10 × 2–4 μm. Edible and quite good.

AUREOBOLETUS

4 Aureoboletus (=Boletus) **cramesinus:** *cap viscid, clay-pink, tubes and pores vivid golden yellow*

In deciduous woods. Autumn. Rare to uncommon. Cap 2–5 cm (1–2 in), convex, smooth and viscid-tacky, often wrinkled; pale pinkish-clay to coral or strawberry-pink, with darker streaks. Stem rather slender, tapered downwards, rooting, pale yellow with reddish lower half. Spores ochre, subfusiform, 11–15 × 4–6 μm. Edible. This unfortunately rare species is easily recognized.

STROBILOMYCES

1 *Strobilomyces floccopus* (Old Man of the Woods, Pine-cone Bolete): *dull grey-black colours; coarsely scaly cap and stem; flesh turns reddish*

In deciduous or conifer woods. Autumn. Uncommon in Europe, frequent in North America. Cap 5–10 cm (2–4 in) convex, covered with very thick, often almost pyramidal, scales; dull blackish-grey to sepia especially on tips of scales; flesh where exposed whitish; margin with overhanging fringe. Tubes and pores white then grey. Stem rather tall, cylindrical, grey-black to brownish; with thick scales forming thick sheathing ring at apex. Flesh white then reddish, finally brown. Spores subglobose with distinct reticulum or network on surface, purplish brown, 10–12 × 8–11 µm. Edible but not recommended.

BOLETINUS

2 *Boletinus cavipes:* *cap yellow-tawny to reddish brown, scaly; pores compound; with thick fleshy white ring*

Under larches in southern England. Autumn. Uncommon to rare in Europe, commoner in America. Cap 3–8 cm (1–3 in), convex then expanded with slight umbo; distinctly felty-scaly especially at margin; distinctly yellow in variety *aureus.* Tubes lemon-yellow to greenish yellow. Pores large, compound, like honeycomb, colour as tubes. Stem yellowish brown, darker below ring, rather fragile and hollow (hence *cavipes*). Spores olive-buff, subfusiform, 7–10 × 3–4 µm. Edible but not recommended.

GYROPORUS

3 *Gyroporus castaneus:* *cap and stem velvety; stem hollow; spore print pale yellow; flesh rather hard, brittle*

Usually below oaks. Autumn. Uncommon, mainly southern in distribution. Cap 4–10 cm (1½–4 in), convex then expanded and flattened, minutely velvety then smooth; rich tawny brown to cinnamon or chestnut. Tubes and pores cream to straw-colour, not changing colour when bruised. Stem rather short, stout, minutely velvety, hollow, colour as cap. Spores elliptic, 8–11 × 4–6 µm. Edible and delicious. This rather attractive small species is best recognized by its spore print, stem characters and rather hard, brittle flesh.

4 *Gyroporus cyanescens:* *cap and stem with rough, hard texture; stem hollow; flesh turns deep blue*

On heathy soils below birch or spruce, northern in distribution. Autumn. Uncommon. Cap 4–15 cm (1½–6 in), convex then expanded, dirty whitish to cream, pale straw or buff; texture rough, velvety-scaly, fibrillose, margin shaggy. Tubes and pores white then yellowish to greenish yellow. Stem rather stout, hollow; colour as cap, fibrillose-tomentose, especially below, rather smooth above; surface often cracking to form ring zones. Flesh firm, white, turning immediately deep blue-green to indigo on cutting. Spores pale straw-yellow, elliptic, 9–11 × 4–6 µm. Edible and good.

PORPHYRELLUS

1 *Porphyrellus pseudoscaber* (=*Boletus porphyrosporus*): *sooty-brown, grey, to blackish colours; spore print purple-brown*

In mixed woodlands, mainly deciduous. Autumn. Widespread but rather rare. Cap 5–15 cm (2–6 in), convex then expanded, velvety then smooth. Tubes and pores pale greyish to pinkish buff, more olive-buff with age, bruising blue-green. Stem quite long, equal or tapered upward; colour as cap with paler, olive-flushed base; slightly velvety, then smooth. Flesh whitish to buff, when cut pinkish grey then grey-olive; rather dark blue-green above tubes and at stem apex. Spores subfusiform, 12–16 × 5–7 µm. Apparently edible, but quality doubtful; best avoided. Cut flesh on white paper stains it green.

TYLOPILUS

2 *Tylopilus* (=*Boletus*) *felleus*: *cap ochre-brown; paler stem with raised network; pores pink when mature*

In mixed deciduous woods. Autumn. Frequent. Cap 5–15 cm (2–6 in), convex then expanding, subtomentose then smooth; tawny ochre to reddish brown, occasionally very pale buff. Tubes and pores white, soon salmon-pink. Stem rather stout, paler, more ochre-yellow or buff than cap; darker, brown network over most of length. Flesh white to pinkish, more or less unchanging. Spores clay-pink, subfusiform, 11–15 × 4–5 µm. Not edible, extremely bitter to taste. Sometimes mistaken for the Cep (*Boletus edulis*), but differing in the distinctly pinkish pores and bitter flesh even when raw.

PAXILLACEAE

PAXILLUS

3 *Paxillus involutus* (Brown Roll-rim): *cap ochre-brown, margin inrolled, woolly; gills and flesh bruise reddish brown*

In mixed woodlands, especially under birch and oak. Autumn. Abundant. Cap 5–15 cm (2–6 in), at first convex, rather umbonate with very inrolled, woolly-shaggy margin; soon expanding, depressed, margin inrolled until fully expanded; subtomentose but often viscid after rain; rich reddish ochre to tawny brown, when young clearly olivaceous at margin. Gills decurrent, crowded, soft and often branched. Stem short, tapered downwards, colour as cap. Flesh pale ochre-buff, darkening when cut. Spores ochre-brown, elliptic, 8–10 × 5–6 µm. *Poisonous*; formerly considered edible and indeed usually is after cooking, but it has been proved to have caused severe poisonings, so must be avoided.

4 *Paxillus atrotomentosus*: *very large brown cap; short, stout, velvet-black stem; on conifer stumps and trees*

Autumn. Frequent. Cap and gills rather like a very large (cap 10–30 cm; 4–12 in) version of the previous species but with very stout, black and densely velvety stem. Spores ochre-brown, ovate, 4–7 × 3–5 µm. Not edible.

GOMPHIDIACEAE

CHROOGOMPHUS

1 Chroogomphus (=Gomphidius) **rutilus:** *often rather large; cap with acute umbo and overall red-brown tints; yellow stem base*

Under conifers. Autumn. Frequent. Cap 4–15 cm (1½–6 in), convex then expanded with more or less acute umbo at centre; smooth and often viscid when wet, shiny when dry; brick-red to reddish brown or reddish olive. Gills deeply decurrent, rather thick, widely spaced, branched at base, olivaceous then soon darkening to blackish brown, edge paler. Stem rather tall, cylindrical, flushed vinaceous (greyish wine red) at apex, yellowish to ochre below and with bright yellow at base; with slight ring zone at apex, surface rather floccose-viscid. Flesh reddish salmon in cap, yellower in stem and bright yellow at base. Flesh + ammonia = violet. Spores sepia brown to blackish, subfusiform, 15–22 × 5–7 μm. Edible but not recommended.

GOMPHIDIUS

2 Gomphidius glutinosus: *cap pale greyish violet to brownish, with spots black or brown; stem also blackens; bright yellow base*

Under conifers, especially spruce. Autumn. Widespread but uncommon. Cap 5–12 cm (2–5 in), convex then expanded with broad umbo then becoming rather depressed; smooth and viscid, pale, whitish to purplish grey or brown and often spotted and mottled black, brown or deep purple-brown. Gills deeply decurrent, whitish then grey, brownish black when mature. Stem medium to long, tapered slightly upwards, covered with glutinous veil which is at first attached to the cap but tears to leave thick, black (with spores deposited) ring at apex, often disappearing; lower stem flushed yellow, fading to whitish above and spotted grey-brown to purple-black as on cap. Flesh + ammonia = vinaceous then brick-red. Spores deep blackish brown, subfusiform, 17–20 × 5–6 μm. Edible but not recommended; rather slimy.

3 Gomphidius roseus: *pinkish, small cap and stem*

Below pines, sometimes in troops. Autumn. Occasional. Cap 2–6 cm (1–2½ in), convex then slightly expanded with low, broad umbo, rather wavy and irregular margin, surface very viscid-glutinous; coral to purplish pink then brick-red with age. Gills deeply decurrent, thick, white then grey, finally deep olive-brown. Stem rather short, stoutish, whitish to pink with wine-red tints; viscid below then dry, with white viscid veil leaving a ring zone at apex. Spores brownish black, subfusiform, 15–18 × 5–6 μm. Edible but not recommended. This attractive little species often grows in association with *Suillus bovinus* (see pp. 198–9).

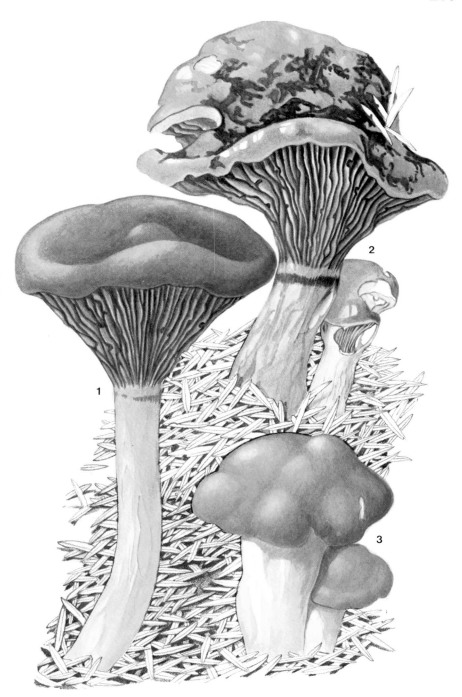

The Bracket fungi and their relatives (Aphyllophorales) form a very large and mixed assemblage varying greatly in shape and size in the various different families. All tend to be rather tough, even hard and woody. The Bracket fungi are familiar, but there are stranger forms such as the coral fungi and the spiny hydnums. The group provides a number of edible species, including the Chanterelle (*Cantharellus cibarius*), eaten by the thousand in Europe. Others, especially brackets, can be serious pests and cause important losses of timber.

LENTINELLACEAE

LENTINELLUS

1 *Lentinellus cochleatus: cap irregularly lobed, funnel-shaped, pale, ochre to date-brown; toothed gills; aniseed smell*

On deciduous wood stumps. Autumn. Uncommon. Cap 3–6 cm (1–2½ in), smooth. Gills decurrent, margins toothed; paler than cap with pink-buff tints. Stem tapered below, fusing into large clumps, colour as cap. Flesh rather tough, often with fragrance of aniseed. Spores white, subglobose, minutely spiny, 5 × 4 μm. Apparently edible but best avoided.

AURISCALPIACEAE

AURISCALPIUM

2 *Auriscalpium vulgare* (Earpick Fungus): *on pine cones; lateral stem; spines below tiny cap*

Autumn. Frequent but difficult to spot as it is extremely well camouflaged. Cap 0.5–1.5 cm, semicircular or kidney-shaped with lateral stem; upper surface minutely hairy, deep brown. Underside with short, pendent spines, brown with greyish flush. Stem slender, very hairy, colour as cap. Spores white, spherical, minutely spiny, 4–5 μm. Not edible. Named after the tiny silver ear-spoons of the last century.

CANTHARELLACEAE

CANTHARELLUS

3 *Cantharellus infundibuliformis* (=tubaeformis): *cap brown, funnel-shaped; "gills" greyish; yellow-orange stem*

In deciduous woods. Autumn. Occasional. More slender than the Chanterelle, with a longer stem. Cap dark brown, "gills" greyish yellow, stem deep yellow to orange (fruit-body entirely yellow-orange in variety *lutescens*). Spores cream, elliptic, 9–11 × 6–9 μm. Edible and delicious.

4 *Cantharellus cibarius* (Chanterelle): *egg-yellow to orange-apricot colours; wrinkled folds instead of gills*

In mixed woods. Autumn. Frequent to common. Cap 3–10 cm (1–4 in), top-shaped then rather funnel-shaped; margin in-rolled, then wavy and irregular; smooth, fleshy. Undersurface bears a series of irregular, blunt gill-like wrinkles and folds, decurrent, paler than cap, often slightly pinkish. Stem short, colour as cap, bright red at base. Flesh thick, fragrance of

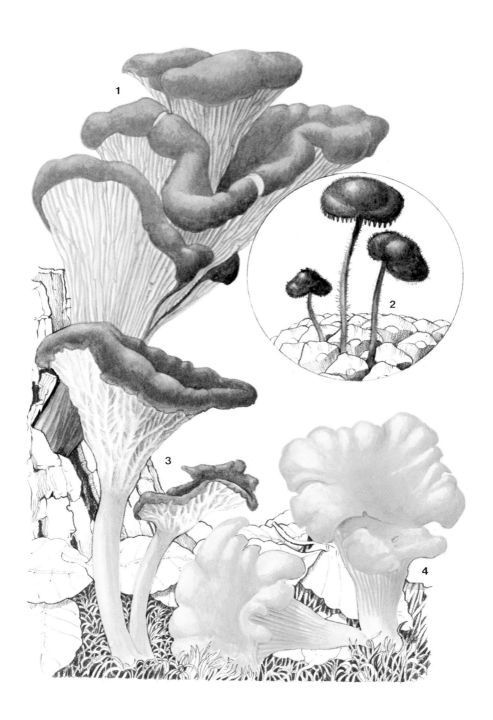

CANTHARELLUS (contd)

apricots when fresh. Spores pale cream, elliptic, 8–10 × 5–6 µm. Delicious; one of the best edible fungi. Compare with False Chanterelle (pp. 134–5).

1 *Cantharellus cinereus: small, clustered fruit-bodies, deep brown, with pale grey "gills"*

In deciduous woods especially near beech stumps. Autumn. Uncommon. Cap 2–5 cm (1–2 in), funnel-shaped, irregular, dark brown. Undersurface with irregular shallow wrinkles or folds, pale grey, frequently branching and forking. Stem short, tapered, often clustered, deep brown. Spores white, ovate, 8–10 × 5–7 µm. Edible and delicious.

CRATERELLUS

2 *Craterellus cornucopioides* (Horn of Plenty, Black Trumpet): *funnel-shaped blackish cap and paler stem; no gills or folds on underside*

In deciduous woods in large numbers below beeches. Autumn. Occasional to frequent. Cap 3–8 cm (1–3 in), funnel-shaped with the margin thin, wavy and irregular; surface rather roughened, minutely scaly; deep brown to black when moist, paler greyish brown when dry. Underside quite smooth to slightly wrinkled, running smoothly into the stem. Stem short, tapered, and quite hollow right down the centre; colour paler than cap. Spores cream, elliptic, 10–11 × 6–7 µm. Edible and good: dries well for grinding as seasoning.

HYDNACEAE

HYDNUM

3 *Hydnum repandum* (Wood Hedgehog): *cap pinkish buff with short pegs or spines underneath; stem white, often to one side*

In mixed deciduous woods. Autumn. Common. Cap 3–10 cm (1–4 in), convex then expanded, thick and fleshy, inrolled at margin, smooth; pinkish yellow to orange-ochre. Gills replaced by short, pendent spines or pegs, densely packed, pinkish buff. Stem rather short, stout, tapered at base, white; often eccentric, giving an uneven, lop-sided appearance. Flesh thick, white, with a pleasant smell, and taste distinctly spicy and slightly bitter. Spores white, elliptic, 7–8 × 6–7 µm. Edible and good, especially when young; best after scalding.

The closely related *H. rufescens* grows under pines and differs in its smaller, deeper reddish brown cap.

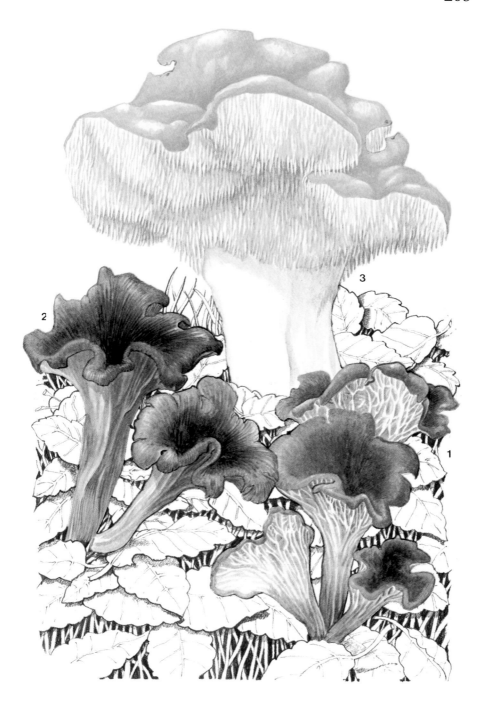

HERICIUM

HERICIACEAE

1 *Hericium coralloides* and *H. ramosum*: *branched white fruit-bodies with pendent spines at tips only in the former species, all along the undersurface in the latter*

In deciduous woods, especially beech and oak. Autumn. Rare to occasional. These two white to yellowish species are very similar and indeed have been confused for many years. Both form large, irregularly branching masses on the trunk of trees or on fallen logs. Size varies from 15–40 cm (6–16 in), and the individual branches have pendent clusters of thin spines at the ends and along the undersurface. In *H. coralloides* the branches are thicker and less finely divided and the spines longer (2–3 cm, 1 in) and clustered mainly at the ends. In *H. ramosum* the branches are finer, the fruit-bodies frequently very large, and the smaller spines (1 cm; ½ in) are spread all along the undersurface of the branches. Both have white spores, more or less globular, in *H. coralloides* 6–8 μm and in *H. ramosum* 3–4 μm. Edible and good after scalding. These spectacular fungi are one of the most beautiful sights of the autumn although unfortunately rare.

Another large species *Creolophus cirrhatus*, on beech, rare, forms very large, 10–30 cm (4–12 in) or more, imbricate, bracket-like fruit-bodies with long, completely white pendent spines, and spines also on the top surface.

2 *Hericium erinaceus*: *rather rounded, with long 6-10 cm spines*

On deciduous trees, especially beech. Autumn. Rare. Another white fungus, but much more globular in shape than the previous species, with the pendent spines very long, 5–8 cm (2–3 in). Spores white, globose, 6–7 μm. Edible and good after scalding.

SARCODON

THELEPHORACEAE

3 *Sarcodon imbricatum*: *thick scaly brown caps with short greyish spines; under conifers*

In pine woods on sandy soils. Autumn. Occasional. Cap 5–20 cm (2–8 in), convex then slightly depressed, funnel-shaped at centre when expanded; dull grey-brown to blackish brown with large, thick, overlapping scales especially at centre. Undersurface with short, decurrent pegs or spines, densely packed, brittle, greyish. Stem short, stout, pale brown. Flesh thick, greyish, rather bitter with age. Spores ochre, globular, minutely spiny, 5–7 μm. Edible when scalded but not recommended.

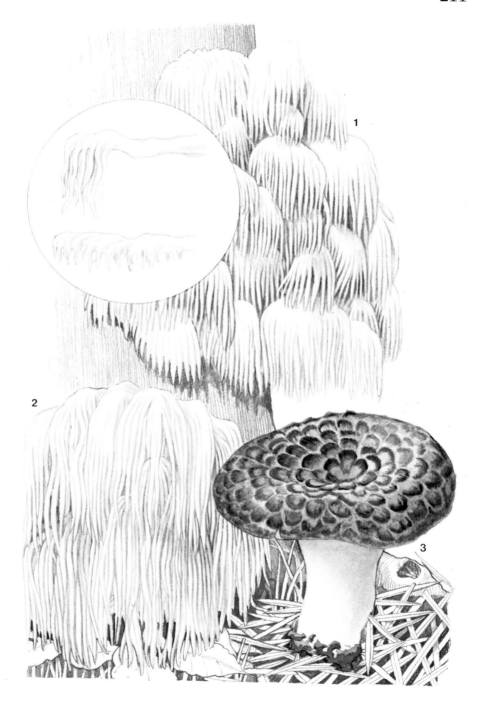

THELEPHORA

1 *Thelephora terrestris* (Earth Fan): *deep brown, fringed, fan-like bodies on soil and leaf-litter*

On soil, and leaf- and needle-litter in pine woods and heathy areas. Autumn. Common. Forms a feathery, fan-shaped structure 3–8 cm (1–3 in) across, often overlapping. Colour a deep umber brown with a violaceous flush, to greyish below. Flesh rather tough, elastic. Spores purple-brown, warted-angular, 8–9 × 6–7 μm. Not of culinary interest. This species is often extremely well camouflaged against the soil, but once spotted it is unmistakable. It has neither gills nor pores, the spores being produced on the smooth to wrinkled surface. *Thelephora palmata*, another species of pine woods, differs in its larger and more coral-like branches, and its foetid smell.

STEREACEAE

STEREUM

2 *Stereum hirsutum*: *greyish yellow brackets; upper surface hairy-woolly, with faint zones*

On deciduous wood, usually fallen logs, branches and twigs. Perennial. Abundant. Forms a thin, tough, semicircular or elongate bracket 3–8 cm (1–3 in) across; yellow-ochre to tawny with slight zones and the upper surface distinctly finely hairy with a greyish flush. Smooth undersurface, rather bright yellow when fresh then fading to buff, not bruising another colour. Spores, white, oblong, 5–7 × 2–4 μm. Not of culinary interest.

S. sanguinolentum is a greyish buff species forming rather more resupinate fruit-bodies which only turn up at the edge to form a bracket; on conifer timber, bruising blood-red.

CHONDROSTEREUM

3 *Chondrostereum* (=*Stereum*) **purpureum** (Silver-leaf Disease Fungus): *small bright purplish brackets; greyish on top*

On various deciduous timber (beeches, poplars and also on fruit trees such as plums) where it causes silver-leaf disease. Autumn. Common. Forms thin, narrow wrinkled brackets 3–8 cm (1–3 in) across; hymenium (fertile layer) on undersurface bright purplish lilac; upper surface slightly hairy-woolly, greyish purple. Spores white, slightly sausage-shaped, 5–9 × 3–4 μm. Not of culinary interest.

HYMENOCHAETACEAE

INONOTUS

4 *Inonotus hispidus* (Shaggy Polypore): *rust-brown, thick fleshy bracket, very shaggy on top*

On the trunks of various deciduous trees, especially ash. Autumn. Frequent. Forms very thick, fleshy and remarkably shaggy brackets, 15–20 cm (6–8 in) across; the upper surface is rich rust- to reddish brown, very densely hairy, darkening to almost black with age. The pores are yellowish at first then reddish, often "weeping" droplets. Flesh deep reddish brown. Spores yellowish brown, subglobose, 8–9 × 7–8 μm. Not of culinary interest. A rather striking species, easily recognized.

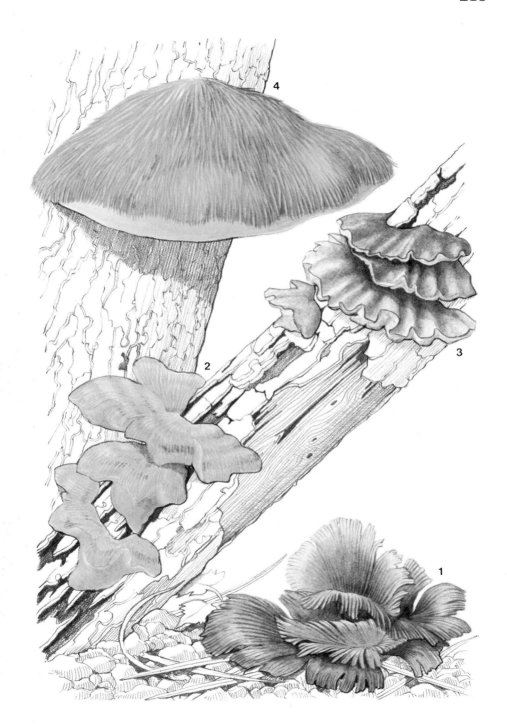

GANODERMA

GANODERMATACEAE

1 *Ganoderma europaeum* (Artist's Fungus): *large thick brown brackets; margin white; rust-coloured spores*

On trunks of deciduous trees, especially beech. Perennial. Abundant, one of the commonest species on beech. Forms large, extremely hard brackets 5–40 cm (6–16 in) across composed of many layers, one added each year. Upper surface smooth, rather irregular, lumpy, rich red-brown with thick white margin. Flesh thick, deep red-brown with strong odour. Pores white to pale brownish cream, bruising red-brown. Spores colour the tree below red-brown; ovate with truncate end, complex channelled spore-wall, 10–11 × 6–7 μm. Far too hard to eat. Called for many years *G. applanatum*, but the true *G. applanatum* is a thinner species with pale cinnamon-coloured flesh. The common name refers to the practice of scratching pictures on the undersurface.

2 *Ganoderma lucidum* (Lacquered Bracket): *bracket with hard, very shiny ("lacquered") reddish surface; lateral stem*

On deciduous trees and logs (causes white rot), always low down. Autumn. Frequent. Cap semicircular to kidney-shaped 10–20 cm (4–8 in), glossy, like lacquer; reddish chestnut to purplish red, margin paler, white or yellow. Pores cream, minute. Stem short to quite long, surface as cap. Spores brown, ovate, with complex chanelled wall, 11–14 × 6–8 μm. Not edible. A spectacular and unmistakable species.

POLYPORACEAE

PIPTOPORUS

3 *Piptoporus betulinus* (Birch Polypore): *thick kidney-shaped, pale brown to whitish, spongy brackets; only on birch*

All year. Abundant. Bracket 10–20 cm (4–8 in) across, with slight or no stem. Upper surface leathery, pale brownish to greyish silver, cracking when very old. Pores (one layer only) pure white, minute when young, soft. Flesh white, sponge-like with many uses (styptic, blotter, tinder etc.). Spores white, sausage-shaped, 5–6 × 1–2 μm. Not edible.

FOMES

4 *Fomes fomentarius* (Tinder Fungus, Hoof Fungus): *thick hoof-like bracket; dull yellow-brown to greyish with paler pores*

On beech in Europe and southern England, but usually on birch in Scotland. Causes white rot. Perennial. Frequent. Brackets 15–30 cm (6–12 in), thick and hoof-shaped, margin blunt; margin pale buff. Flesh pale brown. Pores pale cinnamon with white "bloom". Spores white, elongate, 15–18 × 5–6 μm. Not edible.

(5, see over)

LAETIPORUS

5 *Laetiporus* (=*Polyporus*) *sulphureus* (Sulphur Polypore): *soft-fleshed brackets, bright yellow to pinkish orange overall*

On oaks, chestnuts, and other deciduous trees, occasionally conifers. Autumn. Common. Large often imbricate brackets, irregular but more or less semicircular, 10–40 cm (4–16 in) across. Flesh soft and fragrant. Spores white, pip-shaped, 5–7 × 4–5 μm. Edible and good, best when young.

POLYPORUS

1 *Polyporus squamosus* (Dryad's Saddle, Scaly Polypore): *cap large with darker scales; stem short, black, off-centre*

On stumps, trunks and logs of deciduous timber, causes white rot on living trees. Summer and early autumn. Common. When young a flattened, top-shaped knob, rather like an upturned hoof, soon expanding into a large, almost circular to kidney-shaped bracket 15–35 cm (6–14 in) across; pale yellowish brown, with darker radiating scales flattened onto surface. Pores rather large, angular, decurrent, pale cream. Stem short, thick, hard, white above but deep blackish brown below. Flesh with strong but not unpleasant odour. Spores white, elongate, 10–15 × 4–5 μm. Edible when young and finely shredded, but hardly to be recommended.

2 *Polyporus brumalis* (Winter Polypore): *small round greyish brown cap; whitish central stem*

On fallen deciduous branches and twigs. Late autumn, through winter to spring. Frequent. One of the few brackets to form a cap and stem; cap 3–8 cm (1–3 in) across, rather thin and flattened, smooth, greyish yellow to pale brown. Pores on undersurface white and decurrent, small. Stem thin, tapered at base, whitish and smooth, without a blackish crust at base as in the next species. Spores white, elongate, 5–8 × 1–3 μm. Edible but not recommended.

3 *Polyporus badius* (=*picipes*): *glossy brown funnel-shaped cap; black central stem*

On stumps and fallen wood of deciduous trees. Throughout the year. Occasional. Another species with a cap and stem, the cap round and funnel-shaped, very smooth, glossy, tawny brown to chestnut. Pores small, whitish, decurrent. Stem narrow, tapered at base, covered with a blackish crust (absent in the similar previous species). Spores white, elliptic, 7–8 × 4 μm. Not edible.

The similar *P. varius* is distinguished by its matt, slightly striate cap set off-centre on the thick, black stem.

PHAEOLUS

4 *Phaeolus schweinitzeii*: *at base of conifers; shaggy brown with bright yellow edge*

Autumn. Common. Forms a rounded bracket 10–30 cm (4–12 in) across, sometimes with a thick central basal stump. Cap very rough and densely tomentose, deep reddish brown with bright yellow margin. Pores dull greenish yellow, angular. Stem,

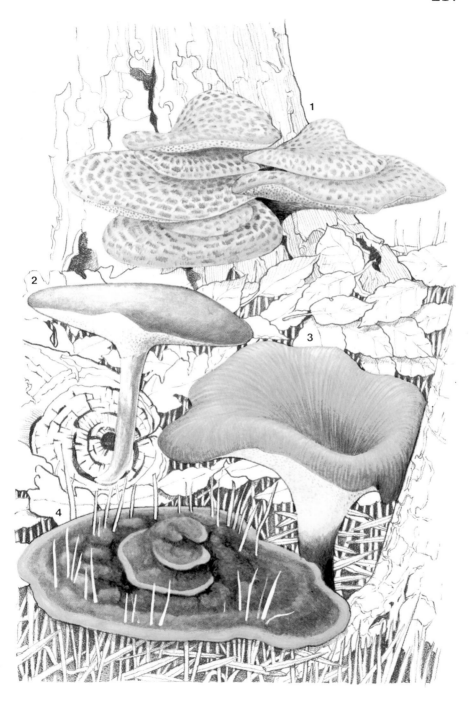

PHAEOLUS (contd)

when present, thick and brown. Spores greenish yellow, elliptic, 8×4 μm. Not edible. This fungus causes serious heart-rot within conifers.

PSEUDOTRAMETES

1 ***Pseudotrametes*** *(=Trametes)* ***gibbosa:*** *thick white brackets, sometimes completely circular; often green with algae*

On stumps of deciduous trees, especially beech. Late autumn. Common. Typically bracket-shaped to, if on the top of a stump, completely circular, 5–12 cm (2–5 in) across. Thick and very tough, fleshy. Upper surface rough and tomentose, knobbly; white to cream, often with algae attached. Pores white, elongated (3–4 times as long as wide). Spores white, elongated, 4–6 $\times$ 2–3 μm. Not edible.

LENZITES

2 ***Lenzites betulina*** *(=Trametes betulinus): pores remarkably thin and elongate, like hardened gills*

On stumps of deciduous trees especially beech and birch. Autumn. Common. Brackets 5–10 cm (2–4 in) across, quite thick, fleshy. Upper surface pale brown, slightly tomentose-hairy, zoned. Pores very elongate, forming thin plate-like or gill-like structures, often branching, but can be strictly poroid. Spores white, elongate, 4–6 $\times$ 2–3 μm. Not edible.

DAEDALEOPSIS

3 ***Daedaleopsis confragosa*** *(=Trametes rubescens)* (Blushing Bracket): *very evenly formed brackets with buff-brown zones; pores white, bruising red*

On deciduous trees and saplings, especially willow. Autumn. Common. Forms perfectly shaped rounded brackets 3–8 cm (1–3 in) across, of medium thickness, tough and fleshy. Upper surface distinctly zoned, pale brownish buff to pinkish or tan-grey; margin paler, to white. Pores quite large, rather elongate, white bruising immediately pinkish red. Brackets usually found as shown opposite, one above the other on thin trunks. Spores white, elongate, slightly sausage-shaped, 10–11 $\times$ 2–3 μm. Not edible.

BJERKANDERA

4 ***Bjerkandera adusta*** *(=Gloeoporus adustus): overlapping greyish brackets, edges soon blackish; pores deep grey to black, minute*

On stumps and logs of deciduous trees. Autumn. Common. Forms masses of rather thin overlapping brackets 4–8 cm (1½–3 in) across, very similar in appearance to *Coriolus versicolor* (pp. 218–19) but differing in the less evenly zoned cap of a dark greyish brown tint, darker at the edge when old but white at first. Spores white, ovate, 4–6 $\times$ 24 μm. Not edible.

MERIPILUS

1 *Meripilus* (=*Polyporus*) *giganteus: very large ochre brackets at base of deciduous trees; pores bruise black in 10 minutes*

Autumn. Common. Irregularly shaped brackets 20–30 cm (8–12 in) across united to form a clump often 200 cm (6 ft) across. Brackets rather thin and quite fleshy, surface leathery, ochre-brown, paler at margin. Pores minute, soft, white bruising blackish. Stem sometimes present although indistinct and merged with cap. Spores white, pip-shaped, 5–6 × 4–5 μm. Edible but not recommended.

GRIFOLA

2 *Grifola frondosa: mass of small brackets from one base; greyish colours, white pores*

At base of deciduous trees especially oaks. Autumn. Uncommon to rare. Forms hundreds of small overlapping brackets clumped to 20–40 cm (8–16 in). Base large, fleshy. Individual brackets 4–6 cm, thin, fleshy, spoon-shaped, greyish on top, white below. Spores white, elliptic, 5–7 × 3–5 μm. Edible, especially the tips, regarded as delicious by many.

DAEDALEA

3 *Daedalea quercina: greyish colours; pores maze-like; usually on oak stumps*

All year. Common. Bracket-shaped, 5–15 cm (2–6 in) across, rather thick-fleshed with hard, corky texture; pale greyish brown, slightly zoned and furrowed. Flesh pale brown. Pores long and irregular, pale greyish brown. Spores white, pip-shaped, 5–7 × 2–3 μm. Not edible.

CORIOLUS

4 *Coriolus* (=*Trametes, Polystictus*) *versicolor* (Varicoloured Bracket): *thin brackets with many-coloured zones; margin white*

On deciduous timber. All year. Abundant. One of the commonest of the smaller brackets, forming semicircular caps which are thin and tough, with clearly defined, silky-velvet, concentric zones of colour, a mixture of browns, yellow-browns, greys, purple, greenish and black, but always with the extreme margin paler, white or cream. Pores cream, small. Spores white, elongate, 6–8 × 2–3 μm. Not edible.

FISTULINACEAE

FISTULINA

5 *Fistulina hepatica:* (Beefsteak, Ox-tongue): *soft, flesh-like, reddish bracket; looks like large tongue*

On oaks. Summer and early autumn. Common. Like a large fleshy, soft tongue, to 50 cm (20 in) across, upper surface soft and spongy, often gelatinous when wet; colour of raw flesh or liver. Flesh red, "bleeds" a reddish juice. Pores pale reddish cream to yellowish, easily separated one from another. Spores pinkish-brown, globose, 4–6 × 3–4 μm. Edible and liked by many (taste rather acidulous, sweet), disliked by others. Causes a brown heart-rot of oak.

CLAVARIACEAE (club fungi)

CLAVARIA

1 _Clavaria argillacea: simple clubs, rather blunt; stem dull yellow, head grey-olive_

On mossy soil or peat in heathy areas. Autumn. Common. A simple, branched club-shaped fruit-body 2–6 cm (1–2½ in) in height with the "head" rounded or blunt, grey-olive, the stem often compressed or grooved, pale yellow-ochre to buff. Flesh yellowish, with taste of tallow. Spores white, elliptic, 10–11 × 5–6 μm. Few if any of the Clavariaceae are of interest as food. Although most are not poisonous, some are very indigestible and laxative in action and all are therefore best avoided.

CLAVULINACEAE (club fungi)

CLAVARIADELPHUS

2 _Clavariadelphus_ (=Clavaria) _pistillaris: large, swollen clubs, ochre-brown and rather wrinkled_

In beech woods. Autumn. Uncommon. This is the largest of the simple club fungi, the club or pestle-shaped fruit-bodies reaching heights of 10–25 cm (4–10 in); the wrinkled surface is yellow-ochre to orange, then reddish-brown. Flesh whitish, slightly bitter. Spores pale yellow, ovate, 11–16 × 6–9 μm. Edible but not recommended.

Other species include the _C. truncatus_, with club also large but flattened on top, more wrinkled, and _C. fistulosus_, with very slender, tall club, rather pointed; both rather uncommon.

CLAVULINOPSIS

3 _Clavulinopsis fusiformis_ (Golden Spindles): _pointed, spindle-shaped simple clubs, bright yellow_

In open grass, heathlands etc. Autumn. Common. Forming tufts of simple, slender, unbranched clubs 5–12 cm (2–5 in) high, spindle-shaped and sharp-pointed, often flattened and irregularly grooved; joined at the base below the soil surface; bright golden yellow. Flesh whitish, with bitter taste. Spores white, globose, 5–7 μm. Not of culinary interest.

The rather similar _Clavulinopsis helvola_ is distinguished by its blunter shape, deeper, orange-yellow coloration and rounded spores with large spines.

CLAVULINA

4 _Clavulina cristata: many-branched, sharp-pointed clubs, white with darker grey-like tips_

In mixed woodlands especially by damp paths and often abundant on streamsides. Autumn. Common. This is a 3–7 cm (1–3 in) high, branching species usually arising from a central thickened stem but ending in fine, pointed branchlets. A colour-form usually regarded as a distinct species, _C. cinerea_, is uniformly grey with blunter, more irregular tips to the branches. Spores white, subglobose, 9–10 × 7–8 μm. Edibility: best avoided.

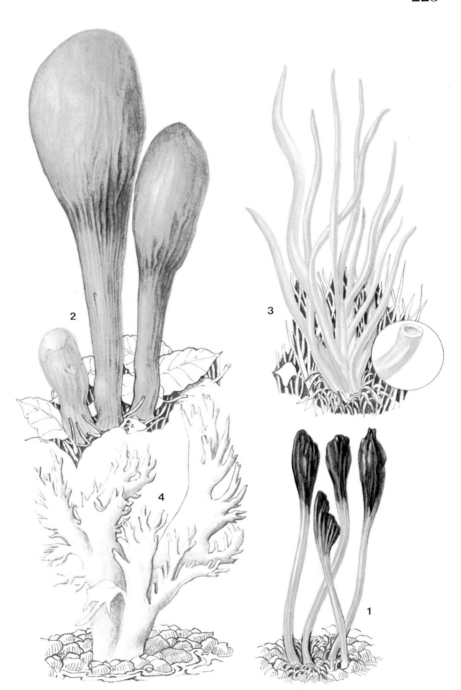

RAMARIA

GOMPHACEAE (coral fungi)

1 ***Ramaria ochraceo-virens*** *(=abietina): branched; deep yellow-ochre colour, bruising green; in conifer woods*

Autumn. Common. Short, thick stem branching into tufts, 3–8 cm (1–3 in) across, tips bluntly pointed. Spores pale brownish, elongate-elliptic, 6–7 × 3–4 µm. *R. flaccida* tougher, more elastic flesh, not bruising green. Neither is edible.

2 ***Ramaria botrytis:*** *robust, large, many-branched; pale tan to ochre with pinkish red tips*

Usually in beech woods. Autumn. Rather uncommon. Thick stem branching into a coral-like head 5–20 cm (2–8 in) across; base whitish. Flesh firm, brittle, white, mild to taste. Spores pale ochre, elongate-elliptic, striate, 12–18 × 4–6 µm. Not edible, can cause upsets. The similar *R. formosa* has a bitter taste, more orange coloration and branch tips yellow.

SPARASSIS

SPARASSOIDACEAE

3 ***Sparassis crispa*** (Cauliflower Fungus): *like a large, pale ochre-brown to tan cauliflower-heart with crisped lobes*

On dead or dying pines, conifer stumps. Autumn. Frequent. Large, 20–40 (50) cm (8–16/20 in) across, composed of hundreds

of flattened curled lobes arising from a central base; texture crisp. Taste mild, odour pleasant. Spores pale, yellowish, pip-shaped, 5–7 × 4–5 μm. The related species (possibly just a form) *S. laminosa* appears to be confined to oaks and has less-branched lobes, longer and more ribbon-like. Both edible and recommended, especially young; may be difficult to clean.

CORTICIACEAE

4 *Merulius tremellosus*: *soft rubbery-gelatinous brackets; furry upper surface, pinkish wrinkled-poroid lower surface*

MERULIUS

On fallen and decaying logs of deciduous trees. Late autumn and winter. Common. Upper surface densely woolly-hairy, whitish. Lower surface with a series of remarkably wrinkled and convoluted shallow pores, pinkish or orange-buff. Spores white, sausage-shaped, 4–5 × 1 μm. Not of culinary interest.

5 *Phlebia radiata*: *flat, spreading, radiately wrinkled; bright pinkish orange against pale bark in early winter*

PHLEBIA

On bark of deciduous trees especially birches, alders. Late autumn and winter. Common. Only the edges are free and at all raised. A spreading sheet or disc of gelatinous tissue, rather wrinkled, with radiating veins or ridges of tissue. Spores white, sausage-shaped, 4–7 × 1–3 μm. No culinary interest.

AURICULARIALES

Although often referred to as jelly fungi the Auriculariales differ from the Tremellales by the basidia not having transverse cross-walls. The texture is also rather more rubbery. Auriculariales and Tremellales are sometimes classified in one order, the Tulasnellales or Tremellales.

HIRNEOLA

1 *Hirneola auricula-judae* (Jew's Ear, Judas' Ear): *soft, reddish brown "ear" on branches, usually elder*

All year. Common. Very like a soft, floppy, velvety-brown ear or cup 3–8 cm (1–3 in) across, the outside being velvety, the inner surface smooth and wrinkled; rubbery, greyish brown to wine-red. Dries bone-hard, but on remoistening resumes its original texture. Spores white, elongate, curved, 16–18 × 6–8 μm. Edible and considered a delicacy in the Far East, where similar species are cultivated. The common name refers to the story that Judas hanged himself on an elder, the "ear" being his returned spirit.

TREMELLALES

All the members of this order have a soft, gelatinous, rubbery or jelly-like consistency and (microscopically) the basidia always divided internally by longitudinal cross-walls or septa (see page 11).

TREMELLA

2 *Tremella mesenterica* (Golden Jelly, Yellow Brain Fungus): *yellow, gelatinous and wrinkled; on deciduous wood*

On fallen branches and stumps of deciduous trees. Late autumn through winter and spring. Common. A gelatinous, slimy fruit-body 3–10 cm (1–4 in across), bright yellow-orange, very irregular and convoluted. Dries bone-hard, deep orange. Spores white, ovate, 7–8 × 5–6 μm. Not of culinary interest.

PSEUDOHYDNUM

3 *Pseudohydnum gelatinosum* (Jelly Tongue): *pearly grey, gelatinous "tongue"; small teeth below*

On stumps and logs of conifers. Autumn. Occasional. Forms a very soft, gelatinous tongue-shaped body 2–6 cm (1–2½ in) across, on a short stem, with the underside covered with short soft pegs or spines; almost translucent, pearly grey. Upper surface slightly hairy. Spores white, ovate, 5–7 × 5 μm. Edible but not recommended.

DACRYMYCETALES

CALOCERA

4 *Calocera viscosa* (Yellow Staghorn Fungus): *yellow-orange, branched, like antlers; on conifer stumps*

Autumn. Common. Forms a slightly sticky, yellow-orange fruit-body shaped like stag's horns, 3–10 cm (1–4 in) in height, rooting in the stump. Spores yellow, ovate, 8–12 × 3–5 μm. The related *C. cornea* grows on deciduous timber and is a simple, unbranched, pointed club, 1–4 cm (½–1½ in) high. Neither species is of culinary interest.

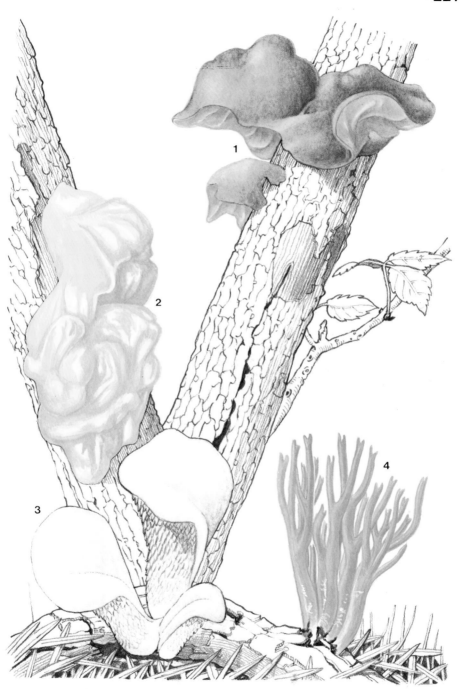

The Gasteromycetes or stomach fungi contain some of the strangest of all fungal organisms and not surprisingly many have been given common names such as stinkhorns, puffballs, earthstars and bird's nest fungi. Despite the remarkable diversity of form, all produce their spores within an outer sterile covering (peridium) and do not discharge them as in the Holobasidiomycetes (pp. 52–227). Instead they usually rely on some external agency to spread the spores, such as insects or mammals, or the physical effects of wind and rain. Despite these common features the Gasteromycetes are now considered as a rather artificial group of several unrelated types which have merely evolved in parallel. They may be divided into the Phallales (stinkhorns), Nidulariales (bird's nest fungi), Lycoperdales (puffballs and earthstars), Sclerodermatales (earthballs) and Hymenogastrales (false truffles).

PHALLUS

1 *Phallus impudicus* var. *togatus*: *like the Common Stinkhorn, but with a fragile, lace-like collar*

Frequent in the U.S.A. but a rare prize in Europe, mainly southern in distribution. Summer and autumn. Height 6–12 cm (2½–5 in); spores 3–5 × 2–3 μm; differs from the Common Stinkhorn only in the remarkable lace-like collarette. Tropical species of the related genus *Dictyophora* have even more expanded collars reaching to the ground like a lace skirt.

2 *Phallus impudicus* (Common Stinkhorn): *unmistakable shape and repulsive odour; begins as an "egg"*

In woodlands, parks and often gardens, usually near deciduous or coniferous stumps. Summer and autumn. Common. Begins as a smooth, egg-like body 5–10 cm (2–4 in) across, buried in the soil, then slowly expanding and protruding above soil level. The outer skin (peridium) splits to reveal a thick jelly-like layer; very rapidly (in a matter of hours) a white spongy stalk surmounted by a globose green head pushes upwards through the jelly, to a final height of 10–20 cm (4–8 in). The greenish black spore mass soon liquefies to produce an obnoxious and extremely penetrating smell which attracts flies to aid in spore-dispersal. The "egg" is connected to very long, tough, white mycelial cords which can often be traced several metres. Spores 3–5 × 2 μm. Sometimes eaten in the unexpanded egg-stage, but usually avoided. This is one of the few fungi possible to hunt by smell! The extremely similar *P. hadriani*, in coastal sand dunes, is rather rare.

MUTINUS

3 *Mutinus caninus* (Dog's Stinkhorn): *cap sharply pointed, not separable; stem rather small, slender, flushed orange*

Woods and gardens. Summer and autumn. Frequent. Smaller (5–10 cm, 2–4 in) than the Common Stinkhorn and without the repulsive odour, differing also in the cap shape and attachment but with the same greenish-black sticky spore-mass, and a variable flush of orange in the stem. Spores 4–5 × 1–3 μm.

CLATHRUS

1 ***Clathrus ruber*** (Cage Fungus): *reddish cage-like body bursting from "egg"; unpleasant foetid odour*

Mainly southern in distribution, frequent in continental Europe and part of America but rare in England. Woods and gardens. Summer and autumn. Beginning as an "egg", like the Stinkhorn (previous page), but soon bursting and expanding to form a spongy, latticed, structure 6–12 cm (2½–5 in) across. Outer surface bright scarlet or pink, inside smeared with sticky, foul-smelling olive-brown slime, the spore mass. Soon attracts flies which eat and disperse the spores. Not considered edible.

CYATHUS

2 ***Cyathus striatus*** (Bird's Nest Fungus): *tiny tall shaggy brown caps containing many small "eggs"; inner edge of cap greyish, strongly striate*

On ground or rotten twigs etc. in mixed woods. Autumn. Uncommon. Small (1 cm; ½ in) cups or cone-shaped fruit-bodies, shaggy brown externally. Greyish fluted inner wall, composed of three thin layers, on which sit 10 or more small egg-like bodies (peridioles) which contain the spores (14–18 × 9–12 μm). When struck by a raindrop the peridioles, which are attached by very thin coiled cords to the inner wall, are splashed out to a distance of several feet and the cord which disengages from the wall helps the spore mass to adhere to surrounding objects (grass etc.). Not of culinary interest. The similar *Cyathus olla* differs in its non-striate, more trumpet-shaped, flaring cones.

CRUCIBULUM

3 ***Crucibulum vulgare*** (Bird's Nest Fungus): *tiny caps with "eggs"; smooth inner surface*

On deciduous or coniferous twigs, clustered. Autumn. Common. Forms small 0.5–1 cm (under ½ in) cup-shaped bodies with yellowish brown, slightly shaggy external surface and smooth, pale yellow, internal peridial layer on which sit 10–12 white "eggs" (peridioles) containing the spores (elliptic, 7–10 × 3–5 μm) (see *Cyathus striatus*, above, for spore-dispersal mechanism). No culinary interest. Another related fungus, *Sphaerobolus stellatus*, is even smaller (0.25–0.5 cm), with star-shaped cups. The inner wall everts to flip the single egg several feet away.

TULOSTOMA

4 ***Tulostoma brumale*** (=*granulosum*): *puffball with long slender stem often buried below sand*

In warm, sandy slopes and dunes. Autumn. Uncommon. Upper chamber 2–3 cm (1 in) across, pale clay in colour. Stem 5–7 cm (2–3 in) long, whitish with small, reddish brown, shaggy scales. Spores pinkish, globose, warty, 4–5 μm; released through an apical pore by wind action and raindrops striking. Not edible. This small and easily overlooked fungus has also been recorded from cracks in old brick walls!

BOVISTA

1 ***Bovista nigrescens:*** *small, whitish puffball without stem, turning deep purple-brown; often detached from ground*

In fields and pastures. Autumn. Frequent. Small to medium sized rounded balls about 3–8 cm (1–3 in) across, without any basal stalk and with the outer skin (peridium) starting whitish at first, then soon darkening with age. The spores are released through an apical opening. The outer peridium layer often peels in irregular patches to expose the very thin, papery inner layer. The whole fungus often becomes detached and rolls around scattering the dark purplish brown spores (globose, smooth, 5–6 μm with long stalk). Edible when young and still firm.

CALVATIA

2 ***Calvatia utriformis*** *(=Lycoperdon perlatum): medium-sized to large; pear-shaped, with hexagonal markings*

In fields and pastures especially on sandy soil. Autumn. Common. A rather large species about 6–10 cm (2½–4 in) across, slightly pear-shaped with a sterile base. Peridium or outer skin is whitish, soon cracking into roughly hexagonal markings; with age turns dirty greyish brown and gradually flakes away to leave only the base. Spores olive-brown and smooth, spherical, 4–5 μm (not warty as in the following species). Edible when young and firm.

3 ***Calvatia excipuliformis*** *(=Lycoperdon saccatum)* (Pestle):
very large, tall, pestle-shaped, with fine granular warts

In woodlands, heaths and occasionally fields. Autumn. Common. An often very large species and taller than other puffballs, reaching a height of up to 20 cm (8 in), greatly resembling a chemist's pestle, with a rounded head surmounting a tall sterile stem. Skin is white and finely granular-warty; the warts soon falling off, leaving smooth skin which eventually flakes away at the apex. Spores olive-brown, warty, globose, 4–5 μm. Edible when young and firm.

4 ***Langermannia gigantea*** *(=Lycoperdon giganteum)* (Giant
Puffball): *unmistakable, often huge; round, smooth and white*

LANGERMANNIA

In fields, pastures and gardens especially by hedges and banks, often reappearing for many years in the same locality. One of the largest fungi of any kind in the world, it has reached diameters of up to 100 cm (3 ft), although the usual size is around 20–50 cm (8–20 in). It forms an irregularly rounded fruit-body, often slightly flattened, with a thick, smooth white skin. Spore mass at first pure white then yellowish. Edible and delicious when young and widely sought after.

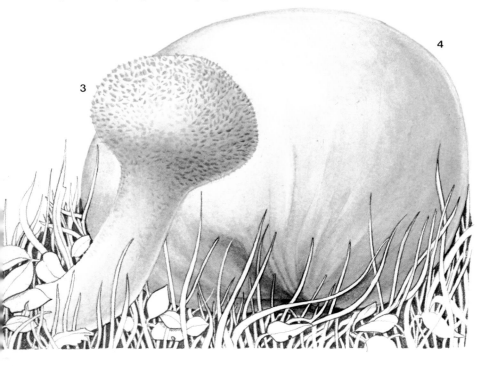

GEASTRUM

1 Geastrum triplex (Earthstar): *large, fat, globose, tulip bulb-like, splitting into 5-7 arms; central round chamber with basal collar*

In leaf litter of beech woods. Autumn. Frequent. When closed forms a rather robust tough brown body which proceeds to split. The outer peridial layer peels back to form five to seven thick, pointed arms often cracking across their inner surface, while the inner peridial layer forms a rounded hollow ball containing the spores. There is an apical pore from which the spores are puffed if struck by a raindrop or dragged out by wind blowing across the aperture. The fungus reaches about 8–12 cm (3–5 in) across the "star". At the base of the puffball-like structure is often a surrounding basal cup or collar. Differs from the similar *G. rufescens* (below) in its larger, more rotund fruit-bodies and its more elongate, prickly spores, 5–6 μm long. Not edible.

2 Geastrum fornicatum (Earthstar): *large 4-armed star with arms bent back and attached to tissue embedded in soil*

In deciduous woods. Autumn. Rare. Outer peridial layer splits to form four abruptly turned-down arms, greyish brown, attached at the tips to a flattened mycelial layer buried in the soil. Central chamber raised on a short stem, also dark brown. Spores finely warty, 3–4 μm. Not edible. A large species, some 5–10 cm (2–4 in) high when open.

3 Geastrum rufescens (=*fimbriatum*) (Earthstar): *reddish brown star-shape with 6-9 arms; puffball at centre without basal cup*

In leaf-litter in deciduous woods, especially oak. Autumn. Occasional, but easily overlooked. At first like a tough, brown onion-like body, then the outer wall (peridium) splitting and folding back to form 6 to 9 pinkish starfish-like arms, 5–10 cm (2–4 in) across. The inner layer of the peridium remains as a puffball structure containing the spores, which are puffed out when raindrops strike, through a distinctly fluted, apical pore. Spores yellowish, prickly, 4–6 μm. Not edible.

4 Geastrum striatum (=*bryantii*) (Earthstar): *small dark brown body splitting into 3-7 arms; central whitish chamber on stalk*

In deciduous woods (especially ash?). Autumn. Uncommon. Much smaller than the other earthstars mentioned here, reaching only 2–5 cm (1–2 in) across the "star". Develops as in the other species, the outer layer splitting to form three to seven dark brown, rather thin arms, but the central chamber is whitish and raised on a narrow stalk with an apical collar. Spores deep brown, warty, 5–6 μm. Not edible.

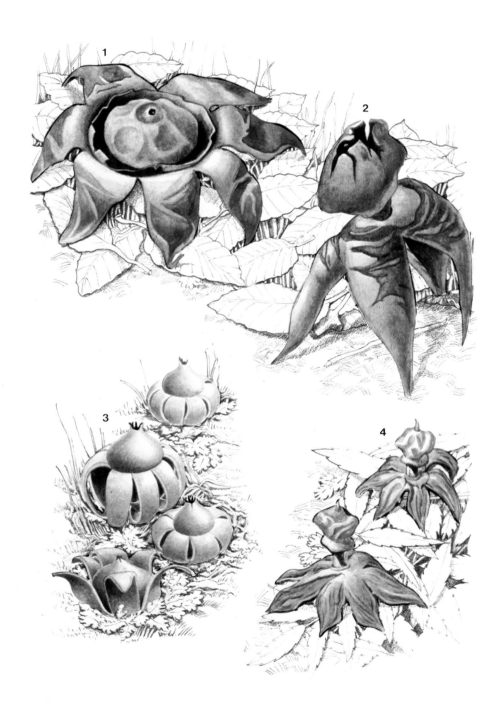

LYCOPERDON

1 *Lycoperdon pyriforme* (Stump Puffball): *always on wood, often in very large numbers*

In large clusters, always on old stumps, logs or buried wood. Summer to autumn. Often abundant. A small pear-shaped species 2–7 cm (1–3 in) high, pale cream-colour to brownish with very fine pointed granules which easily brush off to leave a smooth skin. Stem often has mycelial cords at base penetrating the wood. Spores white then greenish yellow, almost spherical and very finely warty, 4 µm. Edible when young. An easily recognized species and the only European puffball on wood.

2 *Lycoperdon echinatum* (Spiny Puffball, Hedgehog): *brown extremely spiny fruit-body; usually in beech woods*

In deciduous woodlands. Autumn. Very uncommon. Pear-shaped fruit-body, 4–6 cm (1½–2½ in), dark brown and covered with very long brown spines connecting at the tips in small groups. Spines easily break off to leave small rounded scars on the skin. Spore mass deep purple-brown; spores round, warty, 5–6 µm. Not edible. This species is very frequently confused with the common *L. foetidum*, the Common Spring Puffball, another dull brown species, in woods and pastures and with much shorter spines and slightly smaller spores, 3–4.5 µm. *Lycoperdon molle* (= *umbrinum*) is similar but occurs mainly in conifer woods and heaths, with a more distinct basal stem.

3 *Scleroderma verrucosum:* *skin finely warty-scaly, pale brown; rooting stem base; spores olive-brown*

In mixed woods especially on sandy soils. Autumn. Frequent. A smoother, more rounded species than the preceding and with a quite distinct stem-like rooting base which binds together a mass of soil. The skin is pale brownish with fine wart-like scales. Spore mass finally olive-brown, spores 10–14 μm, with spines and ridges. Not edible; *has caused poisoning.*

4 *Scleroderma citrinum* (=*aurantium*) (Common Earthball): *skin thick, scaly, yellowish; central spore mass purple-black*

In mixed woodlands especially birch, near the tree roots. Summer and autumn. Abundant. An irregular rounded fruit-body, 4–8 cm (1½–3 in) across, with a thick, scaly, cracking outer skin, pale yellowish to tawny orange. The solid interior spore mass is white then soon purplish black with a strong, rather unpleasant odour. Spores rounded, 8–13 μm, with a fine reticulate network on the surface. Not edible; cases of *poisoning recorded.*

The cup fungi (Discomycetes) are among the ascomycetes, an enormous worldwide group of several thousand species. Most are microscopic, many are important pests, and all produce their spores within a sac – the ascus (see Introduction). Few ascomycetes are "larger fungi" and most of these are cup fungi. In many ways cup fungi parallel the structures found in the gill fungi, with both toadstool-like and tuberous forms, but the majority produce the typical cup shape.

PEZIZALES

This numerous, often colourful order includes the truly cup-shaped genera and the more complex sponge-like morels and saddle-shaped helvellas. All are rather soft, brittle and fragile, with the spores produced on an exposed layer.

MORCHELLA

1 Morchella conica (Morel): *cap conical, with honeycomb-like pits and ridges aligned vertically*

In fields, pastures and hedgerows, especially on calcareous soils. Spring. Frequent. Height 4–8 cm (1½–3 in). Cap more or less conical, dark grey-brown. Stem short, stout, tapered, brittle and hollow: whitish with a scurfy texture. Spores yellowish, elliptic, 20–24 × 12–14 μm. Edible and delicious.

2 Morchella esculenta (Common Morel): *cap rounded, yellow-brown to reddish; pits and ridges irregular*

In fields, woods and hedgerows. Spring. Common, especially on calcareous soils. Often large, 5–15 cm (2–6 in) high; head rounded, sponge-like; stem white, hollow, often irregular. Spores elliptic, 17–23 × 11–14 μm. Edible, one of the most delicious of all edible fungi. The closely related *M. vulgaris* has a more elliptic and darker (olive-brown to blackish) cap and smaller spores (16–18 × 9–11 μm).

MITROPHORA

3 Mitrophora semilibera: *cap short, conical; cap base free from the stem at its lower margin*

In woods and hedgerows. Spring. Uncommon. 4–10 cm (1½–4 in) high. Cap with longitudinal pits and ridges. Stem tall, thin, white. Spores elliptic, 21–30 × 12–18 μm. Edible.

GYROMITRA

4 Gyromitra esculenta (False Morel): *cap brain-like*

In woods, especially conifers, sometimes in dune slacks; mainly northern. Spring. Uncommon. Distinguished by its brown irregular, convoluted cap, very different from the morels. Stem rather short, stout, whitish. Spores elliptic, 17–23 × 9–12 μm. Not edible; the specific name *G. esculenta* is a misnomer; the False Morel is *deadly under certain conditions*.

HELVELLA

(6, see over)

5 Helvella lacunosa (Black or Grey Helvella): *blackish grey cap, irregularly lobed, folded back over fluted stem*

In woods, reported especially on burnt soil. Autumn. Frequent.